Crimes Against Humanity

Judith Blau is a sociologist who has written extensively on human rights and recently on climate change. In her new book she develops the idea that protecting everyone's human rights and slowing planetary warming are the same goals. It is now clear that the leader of the richest, most powerful country in the world—United States President Donald J. Trump—has set the trigger of destruction by exempting the United States from the international treaty that aims to give the entire planet some reprieve from warming—that is, all countries of the world have entered into an agreement to end reliance on fossil fuels except the United States, which withdrew at the outset of the Trump Administration. Regardless of the U.S. position in the future, the country's emissions are so very extremely high that they will continue to wreck havoc on the entire world. While Blau maintains that President Trump has committed a crime against humanity, even beyond his tenure the book sets the stage for a human rights approach to climate change for the future.

Judith Blau is Professor Emerita in the Sociology Department at the University of North Carolina, Chapel Hill, and was director of the Chapel Hill-Carrboro Human Rights Center, which provided programs and services for refugees and immigrants. She served on the board of the North Carolina American Civil Liberties and was president of Sociologists without Borders, an international nongovernment organization, and the Southern Sociological Society. She formerly edited *Social Forces*, and several of her past publications are about human rights, with others focusing on climate change, architecture, and race.

Crimes Against Humanity

Climate Change and Trump's Legacy of Planetary Destruction

Judith Blau

Routledge
Taylor & Francis Group

NEW YORK AND LONDON

First published 2019
by Routledge
711 Third Avenue, New York, NY 10017

and by Routledge
2 Park Square, Milton Park, Abingdon, Oxon, OX14 4RN

Routledge is an imprint of the Taylor & Francis Group, an informa business

© 2019 Taylor & Francis

The right of Judith Blau to be identified as author of this work
has been asserted by her in accordance with sections 77 and 78
of the Copyright, Designs and Patents Act 1988.

Trademark notice: Product or corporate names may be trademarks
or registered trademarks, and are used only for identification and
explanation without intent to infringe.

Library of Congress Cataloging-in-Publication Data
A catalog record for this book has been requested

ISBN: 978-1-138-31230-2 (hbk)
ISBN: 978-1-138-31268-5 (pbk)
ISBN: 978-0-429-45804-0 (ebk)

Typeset in Bembo and Helvetica Neue
by Florence Production Ltd, Stoodleigh, Devon, UK

For helping us understand why saving the planet and the natural world is urgent, this book is dedicated to:

Richard Delaney (Director, Center for Coastal Studies), Philip Duffy (Director, Woods Hole Research Center), Heather Goldstone (Editor, Science Programs, National Public Radio Cape Cod affiliates), and Bob Prescott (Director, Wellfleet Bay Wildlife Sanctuary).

Contents

Illustrations

FIGURES

TABLES

BOXES

Acknowledgments

Special thanks to local farmers and beekeepers who gave me insights into local practices: Ron Backer, Laura Kelly, Kim Shkapich, and Michael Smith. I would also like to acknowledge and recommend *The Guardian*, which provides outstanding coverage of environmental issues and climate change, with links to science journal articles. A special thanks to a Cape Cod solar company, e-2 solar, not only for installing 21 panels on my roof, but also for raising my interest in renewable energy, and also Pat White of "On Call" who rescued my computer, again and again. For their support and excellent suggestions, I would like especially to thank the editorial team from Routledge—Tyler Bay and Dean Birkenkamp.

Introduction

To slow the pace of planetary heating, by reducing emissions, requires global action—that is, action by all countries and people everywhere. It requires that dictators in the troubled states of North Korea, Eritrea, and Syria participate in this global effort, and, indeed, so far they have. Yet the richest country in the world—the United States of America—has abdicated its responsibility. More specifically, Donald Trump has withdrawn from the Paris Agreement. This does not harm just Americans; it hurts everyone on the globe because the planet is an interdependent system, and the U.S. throwing out carbon dioxide emissions willy-nilly into the atmosphere heats the entire world, not just the U.S. This is a crime against humanity. The consequences are too horrific even to imagine.

The aim of all counties—except, officially, the United States—is to slow warming of the entire planet. Many, perhaps millions, buy and drive an electric car. Even more will install solar panels. Millions of people already have stopped using plastic bags, which saves fish, whales, and even mollusks. Sea walls are being built. Towns and cities are erecting wind turbines. If the entire world's peoples switched to renewable energy by, say, 2040, the rate at which the world is heating would slow. The point I make here is that the challenges are global, requiring deep collaboration with others, and a broadly universal response. We do not have an option.

THE OBJECTIVES OF THIS BOOK

I assume that Trump will either be persuaded to change his mind or he will be impeached. I also assume that Americans will realize that his policies will have catastrophic consequences if they are not abandoned.

In other words, I will highlight what Americans are doing and can be doing, and what other countries are doing. Something to highlight especially is that this is a collaborative project. Another thing to highlight is that this is a project where everyone must cooperate. Yes, the consequences would be catastrophic if Trump succeeded. To prevent that, we all have a huge amount of work to do—build seawalls, install desalinization machines, install solar panels, install wind turbines, and buy electric cars. Everyone must collaborate; everyone must cooperate. One aim, of course, is to ease the transition and settlement of what will be millions of climate refugees. Another is ensuring everyone's human rights. Another is ensuring that everyone has healthy food to eat.

Global warming is not complicated. Basically, it is the "greenhouse effect" whereby gases (carbon dioxide, methane, nitrous oxide, and fluorinated gases become trapped in the atmosphere and heat the planet). The term was coined in 1827 by Joseph Fourier, a French mathematician and physicist, who envisioned that the warming process of the earth acted in the same way as a greenhouse traps heat—a process of visible light and invisible radiation, with the earth's atmosphere acting as the glass barrier.

THE WORLD MOVES AHEAD (WITHOUT THE U.S.)

To achieve global cooperation in order to slow warming, a series of international conferences began in 1979, the most significant of which was COP21 that was held in Paris, from November 30 to December 12, 2015. It was at these meetings that the Paris Agreement was drafted and preliminarily approved by consensus. It is an international treaty crafted with the objective of saving the earth from catastrophic overheating and sparing great human suffering. It entered into force on November 4, 2016, and it establishes that global warming will not exceed 2°C and that a preferred aim is 1.5°C. America was on board and it was approved by President Obama. (As of February 2018, the consensus is that the 2°C objective is insufficient and the aim must be 1.5°C.) That means that in order to slow the heating of the planet, it will be necessary to end all carbon dioxide (CO_2) emissions by 2050, and limit the temperature rise by 1.5°C by the end of the century.[1]

I want to stress the collaborative nature of this project to "save the planet and humankind." Countries formally and voluntarily commit to targets called "Nationally Determined Contributions," which are set by countries to reduce emissions in order to achieve the global goal of achieving net zero emissions before the second half of this century.

According to Article 4, paragraph 2 of the Paris Agreement: "Each Party shall prepare, communicate and maintain successive nationally determined contributions (NDCs) that it intends to achieve. Parties shall pursue domestic mitigation measures, with the aim of achieving the objectives of such contributions."[2]

These NDCs are posted on the Web for public scrutiny, and the objective is for each state to regularly update their NDC so they can meet the objective (along with all other states) of achieving zero emissions by 2050. By January 15, 2018, 165 state parties had submitted their initial plans for reducing emissions.[3] Each state voluntarily describes its own goals in terms of how it will specifically contribute to the worldwide objective of limiting warming to 1.5°. States are encouraged to share strategies and ideas, and as part of a separate process, rich countries assist poorer ones, since rich countries became rich because their early industrial development accompanied spewing out CO_2, much of which still lingers in the atmosphere, continuing to cause global warming. As compensation, the United Nations has set up the Green Climate Fund, which provides assistance to countries like Barbados, or Bangladesh, or Nigeria that do not have the benefits of having industrialized in the 19th century and now bear the costs of converting to renewables.

Just to illustrate what an NDC is, Barbados's initial NDC includes this statement:

> Barbados intends to achieve an economy-wide reduction in GHG [greenhouse gas] emissions of 44% compared to its business as usual (BAU) scenario by 2030. In absolute terms, this translates to a reduction of 23% compared with the baseline year, 2008.[4]

The tragic irony, of course, is that Barbados may never recover from the destructive Hurricane Irma that hit the island in September 2017. These fierce hurricanes are caused by climate change—specifically, the warming of the ocean and sea rise, as I will explain in Chapter 8.

THE WORLD WAS SHOCKED: AMERICA WITHDRAWS

On June 1, 2017, the U.S. President stunned the world when he announced that America would pull out of the Paris Agreement. By November 2017, Syria and Nicaragua had signed on, which means that the U.S. will be the only country in the world that is not party to the Agreement. Indeed, by pulling out of the Agreement and failing to cooperate with the rest of the world, the U.S. will inflict unbelievable harm on the world's peoples by spewing out CO_2 into the atmosphere, thereby accelerating the heating of the planet. (Note, too, that the U.S. is responsible for carbon dioxide that it spewed out in the industrialization era—say, from 1790 to 1860—which still lingers in the environment.)

The reaction to President Trump's announcement was swift and harsh. Scientists and world leaders immediately reacted,[5] and at the climate meetings (COP23) held in Bonn, November 6–17, 2017, world representatives expressed their anger at Donald Trump.[6] Mayors of more than 7,400 cities from around the world vowed to step up their own efforts to help make up what the U.S. would not be doing.[7] In America, there has been an effort to live up to the Paris Agreement regardless of Donald Trump's withdrawal. Already, when the Bonn meetings started in November, 20 U.S. states and 50 of its largest cities, along with more than 600 of the largest businesses had signed up to the "America's pledge" to combat global warming.[8] In addition, there was an alternate delegation representing U.S. states, cities, universities, tribes and businesses at COP23. It was huge and labeled its efforts with the slogan "We Are Still In."[9]

The official U.S. delegation was tiny and gave a session at which they defended coal, consistent with Trump's campaign slogans. This was not only naive, it was ludicrous; coal is a leading cause of global warming. The large audience protested by singing a song to the tune of Lee Greenwood's 1984 *God Bless the U.S.A.* Their words were: "You claim to be an American, but we see right through your greed."[10]

These efforts are extraordinarily helpful, although they are not a substitute for what an entire county can do. Most significantly, the grassroots, even American cities, cannot pass laws or provide large financial assistance to slow climate change—say, by setting a year to end gas-powered cars, as France has done.

DEEP PARTICIPATION, COLLABORATION, AND INCLUSION

Without a strong sense of "We," there is no chance of survival because, after all, we are all united by air, weather, temperature, and seas. And unless we collaborate, we are headed toward catastrophe as the planet heats up, the seas rise, farmlands become arid deserts, cities flood, and fierce storms become common. Happily, over the years since the end of World War II, countries have learned to collaborate and they do so in a myriad of ways. International collaboration, for example, is commonplace in science, trade, monetary policies, the environment, outer space, shipping, sports, the arts, the space station, aviation, mail and communications, agriculture, refugee settlement, the Internet, and criminal justice.

The United Nations is the most important venue and organization for international collaboration of all sorts, and, besides, it is the chief sponsor and facilitator for collaboration involving the climate. Clearly, the climate can only be tackled collaboratively since winds, temperature, air, the oceans are global. The huge and unprecedented problem we face collectively is that the air and oceans are warming at a rate that is extremely dangerous for our health, our food, and our habitats—indeed, it risks our very survival. We must solve this collectively, with alliances involving North Korea, Syria, South Sudan, and other poor countries, and the United States and other rich countries. Anyone who has watched a United Nations General Assembly session on the Internet will immediately understand the point I am making here. (To give an example: "I give the floor to the Gentle Woman from North Korea.")

We must all collectively recognize that the Global South did not cause global warming. Rather, the countries that industrialized in the 18th and 19th centuries bear the most responsibility for the warming that we are experiencing now. Specifically, through the period of industrialization, the factories of the U.S. and Europe spewed out carbon dioxide (CO_2) that was produced by fossil fuels—that is, coal, crude oil and natural gas. Carbon dioxide traps the earth's heat, keeps it from easily escaping, and then it remains in the atmosphere for a very long time. This is why it is imperative for the world to transition quickly away from fossil fuels and to adopt technologies that support renewable energy—notably wind, sun, and tides. There is no official deadline for countries to switch to 100% renewable energy given countries' varied constraints and opportunities, but there is a prevailing consensus that carbon emissions must rapidly decline, beginning in 2020, and that by 2050 renewables must replace fossil fuels.[11] Indeed, this looks doable, assuming that the U.S. rejoins the world community. The costs of renewable energy technology—notably solar and wind—are plummeting throughout the world. Technically and financially, renewables are a snap—at least for rich countries—but in terms of 100% international participation, it is an enormous and unprecedented challenge. Every country must be on board and every country must participate. Third World countries will need financial help (and do remember, that they contributed very little, if any, to global warming). Every single country is on board, except the United States.

To be clear, I recognize that the world is now rife with conflict, wars and aggression. Yet, knowledge of the fate that awaits all of us if we do not cooperate must rivet our full attention and commitment to cooperate. One indication, as I have already said, that this is bound to happen is that *if* (again—*if*) all of the world's countries are on board with the Paris Agreement and have made commitments to cooperate and to end dependence on fossil fuels and to adopt renewable energy. Besides, a goal each country agrees to meet is their own "Nationally Determined Contribution" (NDC). This is a commitment to end emission of their own greenhouse gases by a certain date. Note that there is no coercion but rather a universal understanding of the importance of meeting this commitment.

The 2017 climate meetings themselves expressed what we might call the "we project"—that is, the "we" minus one country. Bonn, Germany, was the conference site, providing abundant resources (that cities of rich countries tend to have), such as venues, transportation, and hotels. But the host and organizer was the country of Fiji, which is a Small Island Developing State, at high risk of being overtaken by the rising sea, and the presiding president of the meetings was Frank Bainimarama, Prime Minister of Fiji. Of great symbolic importance was a huge hand-crafted canoe that Fiji brought to the meetings along with the theme "We are all in the same canoe," as well as the guiding principle *Talanoa*—a Fijian word that expresses a process that is inclusive and participatory, and leads to egalitarian decision-making and promoting the collective good.

Talanoa is consistent with key phrases in the Paris Agreement that highlight cooperation and the common good, and, recognizes the greater responsibilities of countries that chiefly caused global warming. Specifically, countries that industrialized early, from around 1760 to 1830, are most responsible for global warming. This was the era when factories in the U.S. and some European countries spewed carbon dioxide into the atmosphere. Staying in the atmosphere, carbon dioxide continues to trap the heat and is responsible for atmospheric warming, or climate change. This helps us to understand why the Paris Agreement highlights the following:[12]

> In pursuit of the objective of the Convention, and being guided by its principles, including the principle of equity and common but differentiated responsibilities and respective capabilities, in the light of different national circumstances,
>
> Also recognizing the specific needs and special circumstances of developing country Parties, especially those that are particularly vulnerable to the adverse effects of climate change, as provided for in the Convention,
>
> Taking full account of the specific needs and special situations of the least developed countries with regard to funding and transfer of technology,

Acknowledging that climate change is a common concern of
humankind, Parties should, when taking action to address climate
change, respect, promote and consider their respective
obligations on human rights, the right to health, the rights of
indigenous. peoples, local communities, migrants, children,
persons with disabilities and people in vulnerable situations and
the right to development, as well as gender equality,
empowerment of women and intergenerational equity,

THE CAUSE OF GLOBAL WARMING

I explain in lay terms what global warming is in Chapter 1, but here is a
very brief account.[13] It is occurring because greenhouse gases—mainly
carbon dioxide (CO_2)—collect in the atmosphere and absorb sunlight,
causing solar radiation to bounce off the earth's surface. Normally, this
radiation would escape into space, but these gases (pollutants), which can
remain from many years to many centuries in the atmosphere, trap the
heat and cause the planet to get increasingly hotter. In fact, the hottest
years ever recorded were 2014, 2015, and 2016. Besides, global warming
is responsible for sea rise as well as hurricanes, typhoons, cyclones,
acidification of the oceans (absorption of CO_2), desertification, the death
of corals, and toxic pollution. It bears repeating that the biggest culprit of
global warming is CO_2, produced by the burning of fossil fuels. There are
two reasons. First, CO_2 lingers in the atmosphere for a very long time,
causing the rise in temperature, and indirectly causing severe weather
events. Second, CO_2 is absorbed in the oceans causing acidification, which
is responsible for the death of coral reefs, and with destructive effects to
fish and aquaculture. What we are most directly familiar with, of course,
are horrendous weather events, such as heatwaves, droughts, cyclones,
and hurricanes. The connection between these events and climate warming has been examined carefully by scientists and the verdict now is
unequivocally "yes," global warming plays a significant role in most, if
not all of these extreme weather events.

The philosophers would be pleased that there is a plan. It is called the
Paris Agreement. They would be delighted that every country has agreed
to cooperate, except one, the United States—that is, after Barack Obama
signed the Paris Agreement, Donald Trump unsigned it, with withdrawal
in 2020. That makes the U.S. the only country out of 194 that is not a
party to the Paris Agreement. Already, the United States is the second
highest emissions polluter in the world, and it will be the highest once
China becomes party to the treaty and is committed to reduce emissions.

It is important to be clear about the significance of what Trump has done. Pulling out of the treaty means that the U.S., Americans, and American businesses are under no legal obligation to stop polluting, to adopt renewable energy, cooperate with other countries to slow global warming, or assist poor countries to acquire technologies for renewable energy. In other words, Trump's unilateral decision to withdraw from Paris constitutes what the international community calls "A Crime against Humanity"— namely, according to the United Nations,

> In contrast with genocide, crimes against humanity do not need to target a specific group. Instead, the victim of the attack can be any civilian population, regardless of its affiliation or identity. Another important distinction is that in the case of crimes against humanity, it is not necessary to prove that there is an overall specific intent . . . The perpetrator must also act with knowledge of the attack against the civilian population and that his/her action is part of that attack.[14]

The U.S. is so large and emits so much CO_2 that the consequence of Trump's decision is that the country would harm the entire world, not merely Americans. In various chapters—especially Chapter 8—I sketch the ways in which President Donald Trump has already taken specific practical steps to undermine the Paris Agreement. Yet he could get by with this because there is such a high level of climate denial in the U.S., as I describe in the next chapter. Unless there is concerted global action, the world is on track with a 3.2°C increase by the end of the century, which would be at least twice the target in the Paris Agreement.

CONCLUSIONS

From a sociological perspective, I conclude that five things are very important as we ride out the next decades. First, communities, schools and the media need to provide educational opportunities to inform people about climate change so that residents can make smart choices. Second, it is very important that economic inequality is critically addressed and remedied, especially in America where inequality is so extreme. The top 20% of U.S. households own more than 84% of the wealth, and the bottom 40% combine for a paltry 0.3%. Such extreme inequality not only radically undermines society, it impairs people's capabilities of responding to the kinds of emergencies everyone is bound to face. Third, all societies need to adopt generous policies and practices regarding climate refugees, with the expectation that everyone will be at some risk of themselves becoming refugees. Fourth, advancing the overall goals of sustainability and inclusion rather than economic growth will ensure robust participatory engagement and societal robustness that are necessary to deal with the uncertainties that lie ahead. Standing out, and to conclude with, American citizens need a better understanding of the consequences of the U.S.'s withdrawal from the Paris Agreement—specifically, that by not complying with the treaty, the U.S. can harm the entire world by accelerating global warming, which is why I stress that Donald Trump has committed a crime against humanity.

NOTES

1   UN Framework Convention on Climate Change. UNFCCC Newsroom. Available at: http://newsroom.unfccc.int/unfccc-newsroom/finale-cop21/

2   Paris Agreement. Available at: http://unfccc.int/files/essential_background/conven tion/application/pdf/english_paris_agreement.pdf

3   UNFCCC. NDC Registry. Available at: www4.unfccc.int/ndcregistry/Pages/All. aspx

4   "Barbados intended nationally determined contribution communicated to the UNFCCC on September 28, 2015." Available at: www4.unfccc.int/ndcregistry/ PublishedDocuments/Barbados%20First/Barbados%20INDC%20FINAL%20Septe mber%20%2028,%202015.pdf

5   Jonathan Watts and Kate Connolly, "World leaders react after Trump rejects Paris climate deal." *The Guardian*, June 1, 2017. Available at: www.theguardian.com/ environment/2017/jun/01/trump-withdraw-paris-climate-deal-world-leaders-react

6   David Siders, Emily Holden, and Kalina Oroschakoff, "Trump is blasted at climate talks." *Politico*, January 14, 2017. Available at: www.politico.com/story/2017/11/ 14/trump-climate-talks-germany-paris-244903

7   Ian Johnston, "Donald Trump helps spur 7,400 cities around the world into action over climate change." *The Independent*, June 29, 2017. Available at: www. independent.co.uk/environment/donald-trump-climate-change-7400-cities-action- global-warming-environment-paris-agreement-U.S.-a7814256.html

8   America's pledge. Available at: www.americaspledgeonclimate.com/

9   We are still in. Available at: www.wearestillin.com/about

10 Damian Carrington, " 'Tobacco at a cancer summit': Trump coals push savaged at climate conference." *The Guardian*, November 13, 2017. Available at: www. theguardian.com/environment/2017/nov/13/bonn-climate-summit-trump-fossil- fuels-protest. To be sure, just prior to the Bonn meetings U.S. scientists released a report on climate change that was consistent with international science. See U.S. Global Change Research Program, Washington, DC. doi: 10.7930/J0DJ5CTG. Available at: https://science2017.globalchange.gov/. The presentation in Bonn was pitifully out of date and made a mockery of the United States.

11 Christiana Figueres et al., "Three years to safeguard our planet." *Nature*, June 28, 2017. Available at: www.nature.com/news/three-years-to-safeguard-our-climate- 1.22201

12   Paris Agreement. Available at: http://unfccc.int/files/essential_background/conven tion/application/pdf/english_paris_agreement.pdf

13   Then as I became "hooked," I devoured articles and books, and especially recommend the following: Pope Francis, *Encyclical on Climate Change & Inequality* (2015); Al Gore, *An Inconvenient Truth: The Planetary Emergency of Global Warming and What We Can Do About It* (2016); Naomi Klein, *This Changes Everything: Capitalism vs. The Climate* (2014); Mark Lynas, *Six Degrees: Our Future on a Hotter Planet* (2008) and Jeffrey D. Sachs, *The Age of Sustainable Development* (2015).

14   Available at: www.un.org/en/genocideprevention/crimes-against-humanity.html

CHAPTER 1

Heating of the Planet

Is America to Blame?

Every American scientific association has issued a statement proclaiming that climate change—global warming—is occurring. And 98.4% of all the world's scientists also agree *why* the earth is getting warm—namely, because humans are burning fossil fuels, the source of carbon dioxide (CO_2).[1] The solid, uncontestable empirical fact is that unless we limit global warming to less than 2°C (3.6°F), and much more—preferably, 1.5°C (3.4°F)—the world's peoples will be at grave risk—that is, food will be imperiled; many people will starve; habitats will be at risk; some parts of the world will be so hot that they will be uninhabitable; fierce and destructive storms will be a common experience. Global warming is largely caused by carbon dioxide (CO_2) being created when fossil fuels—that is, oil and coal—are burned. The global goal is to end the use of fossil fuels and for everyone—the entire world—to depend entirely on renewables by 2050.

When the *Bulletin of the Atomic Scientists* moved its symbolic Doomsday Clock forward 30 seconds, to two minutes to midnight, on Thursday, January 25, 2018, they cited the U.S. administration and its position on climate change, along with North Korea's nuclear threats, as the reasons for grave alarm.[2] It was not long after that, in February 2018, that the official aim became 1.5°C.[3] Because the U.S. is responsible for a disproportionate share of the world's total carbon dioxide emissions, Trump's withdrawing the U.S. from the Paris Agreement is horrific, and one could say without question, as already argued, he is committing a crime against humanity. It is important to stress that it is not only the U.S. and Americans who will greatly suffer, but the entire world. There are legal grounds that make it a crime—namely, those defined by the ICC.

THE INTERNATIONAL CRIMINAL COURT (ICC)

The Rome Statute, which was adopted in July 1998 and after ratification by 60 countries, entered into force on July 1, 2002. The Rome Treaty sets up an international court to hear crimes of three types: genocide, crimes against humanity, and war crimes (the fourth, crimes of aggression, has not been finalized.)[4] Briefly:

- Genocide is an act committed with intent to partially or wholly destroy a national, ethnic, racial, or religious group.
- Crimes against humanity refer to specific crimes committed in the context of a large-scale attack targeting civilians, regardless of their nationality. These crimes include murder, torture, sexual violence, enslavement, persecution, enforced disappearance.
- War crimes are serious violations of the laws and customs applicable in international or national armed conflict.

The consequences of the U.S. not rejoining Paris to participate with other nations to end reliance on fossil fuels are devastating to the entire world's population as well as to the natural world, and its animals, fish, and all living things. The world's peoples must realize the severity of the act.

Something that is not well understood by the public is the perniciousness of carbon dioxide and the fact that it remains in the atmosphere for a very long time.

CARBON DIOXIDE

China and the U.S. rank, respectively, first and second as the world's top current emitters, as shown in Table 1.1, but it is important to note that China is rapidly becoming a world leader, making a transition from fossil fuels to renewables.[5] However, given the U.S.'s withdrawal from Paris, it will soon overtake China.

It is important to note that carbon dioxide stays in the atmosphere for a very long time, which means that some of the CO_2 emitted during the period of industrialization remains in the atmosphere today. In other words, current emissions underestimate emissions that are actually in the air. Table 1.2 provides information on cumulative CO_2. There is no question that the U.S. has contributed the most to atmospheric CO_2.

Table 1.1 Top 20 Countries: Share of Current CO_2 Emissions, 2015

Country	2015 total carbon dioxide emissions from fuel combustion (million metric tons)	2015 per capita carbon dioxide emissions from fuel combustion (metric tons)
1.	China	9040.74
2.	United States	4997.50
3.	India	2066.01
4.	Russia	1468.99
5.	Japan	1141.58
6.	Germany	729.77
7.	South Korea	585.99
8.	Iran	552.40
9.	Canada	549.23
10.	Saudi Arabia	531.46
11.	Brazil	450.79
12.	Mexico	442.31
13.	Indonesia	441.91
14.	South Africa	427.57
15.	United Kingdom	389.75
16.	Australia	380.93
17.	Italy	330.75
18.	Turkey	317.22
19.	France	290.49
20.	Poland	282.40

Source: Union of Concerned Scientists: www.ucsusa.org/global-warming/science-and-impacts/science/each-countrys-share-of-co2.html#.WmtSaHxG2Uk

HOW DID TRUMP GET BY WITH IT?

It is criminal that President Trump pulled out of the Paris Agreement. We know that all states recognize the collective danger if one country continues to pollute the atmosphere. Indeed, it is catastrophic when a state as large as the U.S. spews out carbon dioxide, polluting the entire globe. Yes, world leaders criticized Trump and so did scientists, but there was

Table 1.2 **Sources of Cumulative Emissions—CO_2—Currently in the Atmosphere (by percentage)**

	%
U.S.	27
All EU countries	25
China	11
Russia	8
Japan	4
India	3
Canada	2
Mexico	1
Brazil	1
Indonesia	1
All others	17

Source: World Resources Institute: Cumulative CO2 Emissions 1850–2011: https://wri.org/blog/2014/11/6-graphs-explain-world%E2%80%99s-top-10-emitters; also see Center for Climate and Energy Solutions: www.c2es.org/facts-figures/cumulative-emissions-1850-2011

Table 1.3 **Percent agree with the Statement "Global climate change is harming/will harm people around the world in the next few years"**

	%
Latin America	95
Europe	86
Africa	85
Asia/Pacific	79
Middle East	70
U.S.	69
Global median	79

Source: Pew Research. "The Politics of Climate Change." Survey conducted May 10 to June 6, 2016: http://assets.pewresearch.org/wp-content/uploads/sites/2/2015/11/Pew-Research-Center-Climate-Change-Report-FINAL-November-5-2015.pdf (pp. 15, 16). Country N's are 1,000: www.pewresearch.org/methodology/international-survey-research/international-methodology/global-attitudes-survey/all-country/2015/

no widespread domestic protest. Why not? An international survey carried out by Pew Research helps to provide the answer.[6]

Table 1.3 presents international figures based on research carried out by Pew as to how much people worry about the immediacy of climate change.

This is remarkable. These comparisons strongly suggest that the reason why Trump could so easily withdraw from the Paris Agreement was relative indifference or denial by a large segment of the American public. Another study carried out by a Yale University climate research center finds that about half of Trump supporters deny or are unsure about climate change.[7] In its study of Americans, Pew Research found striking partisan differences with Democrats far more likely than Republicans concerned about climate change.[8]

In other words, Trump could withdraw from the Paris Agreement because there are enough Americans who do not believe that climate change is taking place. It is important to point out that scientists are clear about what causes global warming, but decisions about how to slow or stop global warming are not made by scientists but instead by publics, politicians, civil society, governments, investors, and businesses. While scientists play a key role in explaining why there is global warming, they cannot make the actual decisions about how to slow it or to ensure protections—that is, to give an example, scientists can analyze and explain why there is sea rise, but they cannot make the decision whether to build seawalls, raise buildings, or whether people should be encouraged to move.

Another decision that Trump made that will harm Americans' ability to respond to climate change was to impose tariffs of 30% on solar panels, which will cost the jobs of about 100,000 installers in America and deter Americans from purchasing them, thereby accelerating the pace that America will contribute to warming.[9]

GREENHOUSE GASES

It has been known for a very long time that human activities warm the air. In 300 BC, Theophrastus carefully documented that the drainage of marshes cooled an area around Thessaly and that the clearing of forests near Philippi warmed the atmosphere. Now we know much more. The warming trend is due to the "greenhouse effect" that results when the atmosphere traps heat radiating from earth toward space. Gases that contribute to the greenhouse effect are those gases in the atmosphere that block heat from escaping and remain semi-permanently in the atmosphere.

Besides water vapor—which does not warm the earth—these include the following:[10]

- Carbon dioxide (CO_2), which is released through natural processes such as respiration and volcano eruptions, and, importantly, through human activities such as deforestation, land use changes, and burning fossil fuels. Humans have increased atmospheric CO_2 concentration by more than a third since the Industrial Revolution began.
- Methane is a hydrocarbon gas produced both through natural sources and human activities, including the decomposition of wastes in landfills, agriculture, and especially rice cultivation, as well as ruminant digestion and manure management associated with domestic livestock. On a molecule-for-molecule basis, methane is a far more potent greenhouse gas than carbon dioxide.
- Nitrous oxide is a powerful greenhouse gas produced by soil cultivation practices, especially the use of commercial and organic fertilizers, fossil fuel combustion, nitric acid production, and biomass burning.
- Chlorofluorocarbons (CFCs) are synthetic compounds entirely of industrial origin used in a number of applications, but now largely regulated in production and release to the atmosphere by international agreement because of their ability to contribute to destruction of the ozone layer.

Yet the greatest risk is from CO_2 because it causes irreversible changes as it continues to accumulate in the atmosphere, remaining there for a very long time.[11] Sources of CO_2 include the following: the burning of solid fuel, such as coal; the burning of liquid fuels, such as oil or petroleum; the burning of gaseous fuels, such as natural gas; industrial gas emissions from manufacturing or processing plants; and deforestation. And, it should be noted, that since 2005 there has been an increase in global carbon dioxide, now standing at 406.9 parts per million. In 2005, this was less than 380 parts.[12] Even if emissions decline, as they have slightly in recent years, carbon dioxide—to repeat—remains in the atmosphere for a very long time. More specifically, between 65 and 80% of CO_2 released into the air dissolves into the ocean over a period of 20–200 years, but the rest remains in the atmosphere for up to several hundreds or thousands of years. In other words, once in the atmosphere, carbon dioxide contributes to climate change forever, at least in any practical term.

Some examples can be cited to illustrate events and phenomena that are unequivocally traced by scientists to climate change. Torrential downpours occurred in the summer of 2017 in Bangladesh that devastated communities, caused food shortages, and were accompanied by fears of epidemics and disease outbreaks.[13] Climate change played a role in the hurricanes that pounded Texas and Florida in September 2017. Unusually warm sea surface temperature contributes to the moisture in the air and warming accelerates wind speed, which spreads wildfires once they start. Scientists also attribute horrific droughts to climate change that occurred in California from 2013 to 2015,[14] and then, in 2017, they traced California wildfires to the conditions created by climate change—namely, dry wood, extremely hot weather, and wind.[15] Yet none of this is new. As I have noted, the "symptoms" of climate change were detected long ago.

Let us take a concrete example. By 2100, Boston is expected to experience a 3–6 foot sea rise if there are no drastic steps taken in the meantime.[16] In response, *Boston Magazine* reported in a special issue that if something were not done, "it will be inevitable that all of the Fenway, the South End, and the Seaport District, as well as parts of Dorchester, South Boston, and downtown—save for a small island near the Common— will be submerged underwater."[17] Clearly, Bostonians, Boston government officials, civil society actors, along with scientists, need to start planning now. What will slow sea rise? Can buildings be raised? Where do seawalls need to be built? Which schools need to be relocated? How can Charles River neighborhoods be protected? The planning is already underway. It is called "Climate Ready Boston." It is comprehensive and collaborative, and includes business leaders, scientists, teachers, politicians, and residents. Note, too, that the federal government is very unlikely to fund Boston's projects or any project that is explicitly related to climate change.

Another way to highlight the gravity of climate change is to cite the 2015 study carried out by the UN Institute for Environment and Human Security that sought to estimate how many environmental migrants—or climate refugees—there will be by 2050. The study concludes by estimating that there will be between 25 million and 1 billion environmental migrants, and some may move within their own countries or across borders, on a permanent or temporary basis.[18] Such a broad estimate is due to the fact that there are many parameters, all of which might vary. For example, will millions of people switch to electric cars that use solar power, thereby reducing the world's dependence on fossil fuels and slowing planetary warming? We can imagine that in countries that are environmentally innovative—say, by supporting electric cars—their citizens will feel more confident pursuing other renewable energy projects.

Clearly, there is reason to be hopeful, so long as we all cooperate. Only if humanity undertakes these challenges of climate warming collaboratively will we come out of this with a greatly improved world —a more egalitarian world. More specifically, how do we do this? Importantly, there are many sources of renewable energy: 1) wind power;

2) solar power; 3) geothermal energy (from the heat of the earth; hot springs); 4) biomass (from wood); 5) biofuels (ethanol and biodiesel); 6) gas converted from landfill; 7) hydropower (dams and small hydropower installations); 8) wave power; and 9) tidal power. (There are a few other proposed sources that are still on drawing boards.) So let's say there are nine sources of renewable energy. What are fossil fuels? How many are there? There are just three: coal, oil, and natural gas. Put in these terms, it seems easy to abandon fossil fuels for good.

Yes, it is true that there are more forms of renewable energy than there are forms of fossil fuel. But are the forms of renewable energy problematic? The blades of wind turbines can trap and kill birds, and some people object to the noise of the blades. Solutions have been found for both problems, including, for example, painting the blades of turbines and not constructing them close to the shore. The efficiency of solar panels, of course, can be improved by mounting them on rotating platforms so they move in synch with the sun. The most intractable problem has been biofuels—namely, the conversion of plants and crops into fuel. While burning biofuels does not contribute to planetary warming, biofuels are produced by burning food, such as sugar cane, corn, maize or wheat— namely, food that people can eat and for that reason is highly objectionable.[19] (However, promising work is being done to explore the possibility of using algae as biofuels.)[20]

Probably the most important thing to flag is greenhouse gases (GHGs), which are the main culprit. These are gases that are produced by burning fossil fuels, and because they trap heat in the atmosphere, they warm the earth. In the U.S., GHG emissions in a typical year include: carbon dioxide or CO_2 (81%), methane (11%), nitrous oxide (6%), and fluorinated gases (3%).[21] Carbon dioxide (CO_2) is mainly produced by burning fossil fuels and is partially removed from the atmosphere (or "sequestered") when it is absorbed by plants and trees. (Some 87% of all human-produced carbon dioxide emissions come from burning fossil fuels like coal, natural gas, and oil. The remainder results from the clearing of forests and other land use changes (9%), as well as some industrial processes such as cement manufacturing (4%).)

Methane (CH_4) is emitted during the production and transport of coal, natural gas, and oil. Methane emissions also come from livestock flatulence, other agricultural practices, and from the decay of organic waste in landfills. Nitrous oxide (N_2O) is emitted during agricultural activities and is produced by fertilizers, as well as during combustion of fossil fuels and solid waste. Fluorinated gases include hydrofluorocarbons, perfluoro-carbons, sulfur hexafluoride, and nitrogen trifluoride, and extremely power-ful greenhouse gases that are emitted from a variety of industrial processes.

There is agreement among scientists that emissions from fossil fuels must be reduced to zero by 2050. It would be calamitous were this not to happen. Precise estimates of what will happen *when* are not possible since many events are triggered by other events, and although scientists are certain about the events and the processes themselves, they cannot predict the magnitude and timing of these events and processes. For example, scientists know that when (not if) the Greenland Ice Sheet melts, the sea level will rise about 6 meters (20 feet). When the Antarctic Ice Sheet melts, the sea level will rise by about 60 meters (200 feet). That is a lot. But *when* each happens depends on, first, when and how the ice sheet breaks loose from its land attachment and, second, the rate that water flows under the ice sheet, melting it from below. The consequences are unimaginable. Around the globe there are large population concentrations in cities situated on oceans. Major cities—New York, Tokyo, Shanghi, Los Angeles, Miami, New Orleans—will be under water. It is important to stress that the cause of climate change can be explained and understood clearly, in nontechical terms. It bears repeating that the reason is that the burning of fossil fuels results in the earth warming at an unacceptable rate and this is because of the release and build-up of greenhouse gases that trap heat in the atmosphere. Of these greenhouse gases, carbon dioxide

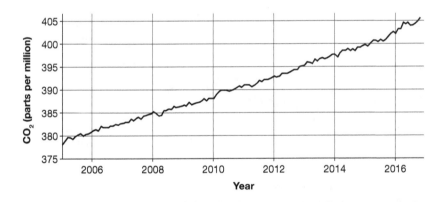

Figure 1.1 Carbon Dioxide (CO_2) Levels: Parts per Million (ppm)

Note: CO_2 is a heat-trapping (greenhouse) gas that is released through human activities, such as deforestation and burning fossil fuels. 350 ppm has been defined as the safe and acceptable level of CO_2 and passing the 360 ppm and now the 400 ppm mark spells danger, especially for marine life, and fuels fears of out-of-control and irreversible warming. Between 65% and 80% of CO_2 released into the air dissolves into the ocean over a period of 20–200 years, causing acidification that is particularly harmful to coral and shellfish. The rest remains for as much as several hundreds of thousands of years. The concentration in the atmosphere is currently 40% higher than it was when industrialization began.

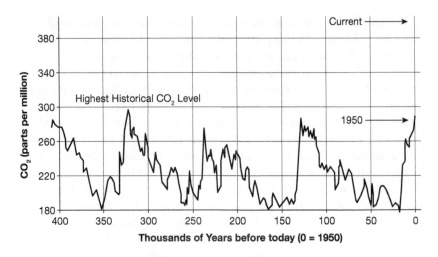

Figure 1.2 Historical CO_2 Level in Parts per Million (ppm)

Note: The dotted line shows the highest historical levels of CO_2 with dramatic contemporary increase detected in 1950.

Source: climate.nasa.gov

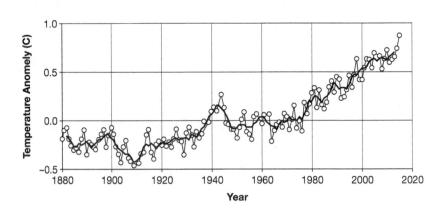

Figure 1.3 Global Temperature

Note: The graph shows the change in global surface temperature relative to 1951–1980. The ten warmest years in this 134-year record all have occurred since 2000, with the exception of 1998.

is the major culprit. Again, to repeat, burning fossil fuels—coal, oil, and naural gas—is what produces CO_2.

It is useful to present graphs that summarize the basis for what are probably the best reasons why there is so much concern about planetary warming. These graphs were prepared by the Earth Science

Communication Team at the National Aeronautics Space Administration's (NASA's) Jet Propulsion Laboratory, in collaboration with a variety of other scientific organizations, including the National Oceanic and Atmospheric Administration and NASA's Goddard Institute for Space Studies.[22] These graphs show past trajectories (up to the present), but do not incorporate the consequences going forward from 2016 of halting or slowing emissions of GHG.

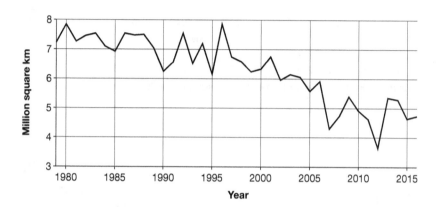

Figure 1.4 Arctic Sea Ice Minimum: Average September Extent

Note: Arctic sea ice reaches its minimum each September. Relative to the 1981 to 2010 average, sea ice declined at a rate of 13.3% per decade. The 2012 sea ice extent is the lowest on record.

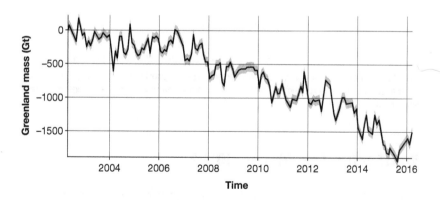

Figure 1.5 Antarctica Mass Variation Since 2002

Note: The continent of Antarctica has been losing an estimated 134 gigatonnes of ice per year

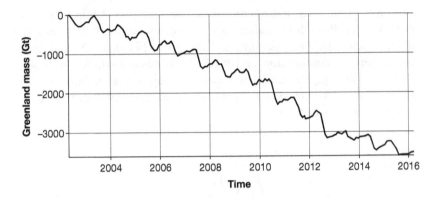

Figure 1.6 Greenland Mass Variation Since 2002

Note: The Greenland ice sheet has been losing an estimated 287 gigatonnes per year.

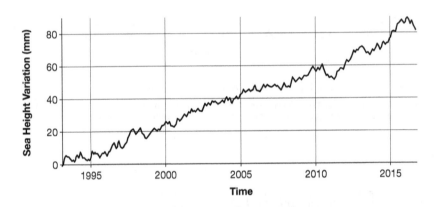

Figure 1.7 Change in Sea Level (Last Measured July 2016): Satellite Data

Note: Sea level rise is caused primarily by two factors related to global warming: the added water from melting land ice and the expansion of sea water as it warms. The chart tracks the change in sea level since 1993.

WHAT DOES THE FUTURE HOLD?

Because these graphs do not include future projections, they do not help us to understand what the world will be like, say, 50 years from now or at the end of the century. Before summarizing such projections it is important to stress that great efforts are being made all around the world to steer clear of the catastrophes that are in store for us were we to do nothing. The means to do so are straightforward—namely, the world's peoples must abandon fossil fuels, since they are the cause of greenhouse gases. Table 1.4 summarizes what scientists predict will happen if nothing is done to end reliance on fossil fuels. It should be noted that scientists are exceedingly cautious and conservative when they make projections, and is very likely that these projections are way too conservative.

The countries that industrialized first, in the 19th and 20th centuries, bear most of the responsibility for the cumulative emissions that now linger in the air. As Table 1.2 shows, the U.S. is the leading source of current atmospheric emissions, followed by all the EU countries combined. This makes it clear why countries all around the globe were angry that the U.S. President withdrew from the Paris Agreement, which frees U.S. industries, government agencies, municipalities, and households to use fossil fuels, spew out carbon dioxide, and continue to heat up the planet.

HOW WE GOT TO WHERE WE ARE TODAY

The significance of country contributions to cumulative emissions is deeply troubling. The U.S. and European countries were advantaged by early industrialization—the U.S. by cheap, enslaved labor, and European countries by the exploitation of colonial resources. The U.S. especially, but also European countries, gained immensely from early industrialization that was possible from the exploitation of fossil fuels. Now it is imperative to abandon these fossil fuels and switch to renewable energy, which is entirely feasible for the rich countries that benefited from fossil fuels, but very expensive for developing countries. With the global transition to renewable energy, it is important that rich countries contribute to the Green Climate Fund to enable developing countries to acquire renewable energy. For this reason, the Paris Agreement clarifies "differentiated responsibilities"—namely, that rich countries will shoulder more of the costs than developing countries. President Trump has ordered that the U.S. should stop payment to the fund.

Below, I will give the background leading up to the Paris Agreement and in the next chapter describe the significance of the Paris Agreement, all of which is necessary as a background to understanding how climate change is likely to affect our lives, our food, and, yes, our chances of our losing our homes and becoming climate migrants. It is important to trace the history, even if briefly, of the scientific background to clarify why and how the planet is getting warmer.

Table 1.4 Climate Change Projections if Nothing is Done

Temperature

- It is very likely that hot extremes and heatwaves events will become more frequent.[a]

- Increases in average global temperatures are estimated to reach 3.6– 4°C (39°F).[b]

- Global average temperature is expected to warm at least twice as much in the next 100 years as it has during the last 100 years.[b]

- Ground-level air temperatures are expected to continue to warm more rapidly over land than oceans.[b]

- Parts of the Persian Gulf may be uninhabitable by 2100.[a]

Precipitation and Storms

- Global average annual precipitation through the end of the century is expected to increase, although changes in the amount and intensity of precipitation will vary significantly by region.[b]

- The intensity of precipitation events is likely to increase. This will be particularly pronounced in tropical and high-latitude regions, which are also expected to experience overall increases in the frequency of precipitation.[b]

- The strength of the winds associated with tropical storms is likely to increase. The amount of precipitation falling in tropical storms is also likely to increase.[b]

- It is certain that future tropical cyclones (typhoons and hurricanes) will become more intense, with larger peak wind speeds and more heavy precipitation associated with ongoing increases of tropical sea surface temperatures.[a]

- Extra-tropical storm tracks are projected to move poleward, with consequent changes in wind, precipitation and temperature patterns, continuing the broad pattern of observed trends over the last half-century.[a]

Ice, Snowpack, and Permafrost

- Snow cover is projected to contract. Widespread increases in thaw depth are projected over most permafrost regions.[a]

- Sea ice is projected to shrink in both the Arctic and Antarctic. In some projections, Arctic late-summer sea ice disappears almost entirely by the latter part of the 21st century.[a]

- For every 1.1°C (2°F) of warming, models project about a 15% decrease in the extent of annually averaged Arctic sea ice and a 25% decrease in the area covered by Arctic sea ice at the end of summer.[c]

- The coastal sections of the Greenland and Antarctic ice sheets are expected to continue to melt or slide into the ocean. If the rate of this ice melting increases in the 21st century, the ice sheets could add significantly to global sea-level rise.[c]

- Glaciers are expected to continue to decrease in size. The rate of melting is expected to continue to increase, which will contribute to sea-level rise.[d]

continued . . .

Table 1.4 Continued

The Ocean, Shellfish, Food, Agriculture

- Besides warming the atmosphere, CO_2 is changing the chemistry of the ocean by increasing its acidity. The result is the death of coral reefs and the erosion of the shells of oysters, clams, and other shellfish.[e]

- Climate change is making increases in agricultural productivity even harder to achieve. These impacts are increasing the risk of hunger and the breakdown of food systems. Without considerable efforts made to improve people's climate resilience, it has been estimated that the risk of hunger and malnutrition could increase by up to 20% by 2050.[f]

- Climate change already affects agriculture and food security; without urgent action, millions more people will be at risk of hunger and poverty.[g]

- Increasing ocean temperatures have significantly affected marine life, inducing shifts in distribution and changes in abundance. Climate change alters the distribution of suitable habitats, forcing fish and other organisms to move to a more favorable area of their range.[h]

- 800 million people are hungry today, and population growth will require food production to increase by about 60% by 2050.[i]

Sources:

a) Intergovernmental Panel on Climate Change. *Fourth Assessment Report* (2007). Available at: www.ipcc.ch/publications_and_data/ar4/wg1/en/spmsspm-projections-of.html

b) Intergovernmental Panel on Climate Change. *Fifth Assessment Report* (2013) Available at: www.ipcc.ch/report/ar5/wg1/

c) National Research Council (2011). *Climate Stabilization Targets*. Available at: www.nap.edu/read/12877/chapter/1#x

d) United States Global Change Research Program (2014) *Climate Change Impacts in the United States*. Available at: http://s3.amazonaws.com/nca2014/high/NCA3_Climate_Change_Impacts_in_the_United%20States_HighRes.pdf

e) Environmental Protection Agency, Effect of Ocean and Coastal Acidification on Ecosystems. Available at: www.epa.gov/ocean-acidification/effects-ocean-and-coastal-acidification-ecosystems

f) World Food Program: Climate Change. Available at: www.wfp.org/climate-change

g) Food and Agriculture Organization: FAO's Work on Climate Change. Available at: www.fao.org/3/a-i6273e.pdf

h) National Oceanic and Atmospheric Administration. Distribution of Fish on the Northeast U.S. Shelf Influenced by both Fishing and Climate. Available at: www.nefsc.noaa.gov/press_release/pr2014/scispot/ss1414/

i) Food and Agriculture Organization. Climate Change. Available at: www.fao.org/climate-change/our-work/what-we-do/en/

A BRIEF ACCOUNT OF THE HISTORY OF SCIENTIFIC EFFORTS TO TRACK CLIMATE CHANGE

Beginning as early as the late 19th century, scientists contended that emissions of greenhouse gases (especially carbon dioxide) warm the atmosphere, but they lacked the necessary equipment to carry out precise measurements.

Computer modeling and observational techniques improved over the next decades, and in 1979 scientists organized the First World Climate Conference. Convened by the UN World Meteorological Organization (UNWMO), it was held on February, 12–23 in Geneva, and was attended by scientists from a wide range of disciplines. In addition to the main plenary sessions, the conference organized four working groups to look into climate data, identify climate topics, carry out integrated impact studies, and do research on climate variability and change. It called on the world's governments "to foresee and prevent potential manmade changes in the climate that might be adverse to humanity." The conference led to the establishment of the World Climate Programme and to the creation of the Intergovernmental Panel on Climate Change (IPCC) by UNWMO and the UN Environmental Programme (UNEP) in 1988.

A number of intergovernmental conferences were held in the late 1980s, raising international concern about planetary warming. These included the Villah Conference in October 1985, the 1987 Montreal Protocol on Substances that Deplete the Ozone Layer (which was amended in Kigali in October 2016), the Toronto Conference in June 1988, the Ottawa Conference and the Tata Conference in February 1989, the Hague Conference and Declaration in March 1989, the Noordwijk Ministerial Conference in November 1989, the Cairo Compact in December 1989, and the Bergen Conference in 1990.

The Second Climate Conference was held on October 29 to November 7, 1990, again in Geneva. It was an important step toward a global climate treaty. The main task of the conference was to review the Intergovernmental Panel on Climate Change (IPCC), the first international assessment report on climate change. The scientists and technology experts at the conference issued a strong statement highlighting the risks of climate change. The conference issued a Ministerial Declaration only after hard bargaining over a number of difficult issues. Yet the Declaration disappointed many of the participating scientists as well as some observers who felt that it did not offer a high enough level of commitment. Eventually, however, developments at the conference led to the establishment of the United Nations Framework Convention on Climate Change (UNFCCC), of which the Kyoto Protocol is a part, which aims to curb emissions, and to the establishment of the Global Climate Observing System (GCOS), a global observing system for climate and climate-related phenomena.

In December 1990, following the Second Climate Conference, the UN General Assembly approved the start of treaty negotiations. In 1992, the treaty, "part of the UN Framework Convention on Climate Change (UNFCCC)," was signed by 154 states (plus the EU) at Rio de Janeiro and now by 197.[23] The U.S. ratified the treaty.[24] The Convention—the Rio Declaration—entered into force on March 21, 1994, and in September parties started submitting reports detailing strategies to deal with climate change. One guiding principle of the Convention was that the richest countries commit to returning gas emissions to 1990 levels by 2000 and they take the lead in reducing emissions.

Trump's decision to withdraw from the Paris Agreement contradicts the U.S.'s ratification of an earlier climate treaty, specifically, the 1992 Rio Declaration, with its insistence on: 1) sustainability as a worldwide objective; 2) "differentiated responsibilities," which is to say rich, developed countries have responsibilities to help developing ones (that are largely not responsible for climate warming); and 3) protecting human rights, especially the rights and well-being of vulnerable people.[25] Besides, the Paris Agreement, like the Rio Declaration, is part of the UN Framework Convention on Climate Change (UNFCCC).

CONFERENCE OF THE PARTIES (COP)

Any Conference of the Parties ("COP") is a governing body of an international convention, and includes all parties to the convention. Most well known is the UN Framework Convention on Climate Change (UNFCCC).[26] Because it is important to see the progression in the scope and themes of the annual (sometimes biannual) UNFCCC–COP meetings, as well as the diversity of meeting sites, I give the date and city of each meeting. What is most important is the 2015 Paris Agreement, or COP21. It lays the framework for dealing with climate change until 2100.

COP1 was held in Berlin in 1995, in response to the growing concern among the world's scientists about climate change. It grappled with resolving equity for developing nations. (As already noted, the 1995 COP1 conference followed a series of international climate meetings, beginning in 1979 with the First World Climate Conference.) COP2 was held in Paris in December 1995, and led to the Second Assessment Report, written and reviewed by some 2,000 scientists from all over the world. The report concluded that there was evidence that there is "a discernible human influence on climate change."

The Kyoto Protocol, which sets standards for emissions to reduce greenhouse gases, was adopted at COP3 in Kyoto, where some 10,000 delegates attended. COP4 was held in Buenos Aires in 1998 (to develop implementation of the Kyoto Protocol) and COP5 was a technical meeting, held in Bonn in 1999. COP6, held in The Hague in 2000, ultimately collapsed because of disagreements between the U.S. and some European countries over satisfying a major portion of U.S. emissions by allowing credits for "agricultural sinks." Yet the U.S. had withdrawn from the Kyoto Protocol and therefore was not a member of COP, but, consistent with the rules, took the role of mediator at the meetings. And likewise, the U.S. was not a party at the COP7 meetings held in Marrakech in 2001. At COP8 (held in Delhi) emission standards were reviewed and updated. At COP9, held in Milan in 2003, an Adaption Fund was approved that would assist poor countries better adapt to climate change. In Buenos Aires at the COP10 meetings, a main focus was emerging economies and developing countries.

COP11 led to the Montreal Action Plan, "to negotiate deeper cuts in greenhouse emissions." COP12 took place in 2006 in Nairobi; COP13 (2007, Bali), and COP14 (2008, Poznan, Poland) continued efforts to strengthen the financing of poor countries as well as to carry out negotiations for a successor to Kyoto.

The World Climate Conference-3 (WCC-3) was held in Geneva, Switzerland, August 31 to September 4, 2009. Its focus was on climate predictions and creating information formats for decision-making at the seasonal to multi-decadal meetings of experts. The goal was to create a global framework that would link scientific advances, specifically with regard to climate predictions, and the needs of their users for decision-making to better cope with changing conditions. Key users of climate predictions include food producers, water managers, energy developers and managers, public health workers, national planners, tourism managers and others, as well as society at large. Participants in WCC-3 included many users, scientists, as well as high-level policy-makers. The Conference also aimed to increase commitment to, and advancements in, climate observations and monitoring to better provide climate information and services worldwide to improve public safety and well-being. COP15 was held in Copenhagen in December 2009 and included negotiations on a framework for a longer time commitment. It included ministers and officials from 192 countries, although the United States still refused to go along with Kyoto.

At COP16, held in Cancún in 2010, the parties agreed that "climate change represents an urgent and potentially irreversible threat to human societies and the planet and thus requires to be urgently addressed by all Parties." Participants agreed that the goal would be a maximum of 2°C. In Durban (2011), at COP17, the parties agreed to begin work on a legally binding treaty, and to advance the Green Climate Fund that would assist poor countries to battle climate change. The Green Climate Fund is part of the Paris Agreement and to illustrate its importance, in 2015 India made a commitment to reduce emissions by 33–35% by 2030, compared to

2005 levels. It also pledged to achieve 40% of electricity from non-fossil fuel-based resources by 2030.[27] But India will need financial assistance and this will be provided through the Green Climate Fund, into which rich countries contribute.[28]

COP18 met in Doha (2012) and reached decisions about compensation to poor countries that experienced grave harms from climatic events. COP19 was held in Warsaw in 2013, and was devoted to planning for COP21 to be held in 2015. At COP20 in Lima (2014), pledges were made for the Green Climate Fund and there was further preparation for COP21. COP22 was held in Marrakesh in November 2016, and COP23 was brilliantly co-organized by Bonn and Fiji. With Fiji's participation, the plight of small island states was highlighted.

CONCLUSIONS

In this chapter I have sketched the basic principles of climate change—namely, how the emissions from burning fossil fuels trap heat to cause rapid warming of the atmosphere and planet, and why it is imperative to switch over to forms of renewable energy. Scientists are confident that if every country and all the world's peoples cooperate to ensure the end of fossil fuels by 2030, we will be able to minimize the harms. Yet one thing still remains unknown, and that is whether or not oceans will lose their capacity to absorb carbon dioxide. If they do, geoengineering solutions are likely. Some possibilities are discussed in Chapter 10. Climate change (sea rise, extreme weather events, extinctions, etc.) may be the most straightforward set of problems with which we must contend. Exceedingly complicated are racial and economic inequalities that impede cooperation, and, besides, are unfair. The U.S. is an outlier with its exceedingly high inequality—so high, in fact, that the United Nations is currently investigating this, with a special focus on human rights.[29]

It should be noted that in eight short years, between 2007 and 2015, actual global reliance on renewables just about doubled, from 5.2% to 10.3%, and more than half of this occurred in developing countries.[30] Moreover, the World Economic Forum reported in 2017 that for 30 countries the costs of renewable energy—wind and solar—for generating electricity were less than those of fossil fuels.[31] One thing is certain: that collaboration, cooperation and trust are required. We can only save planet earth if we work together as partners. The 2015 Paris Agreement was forged as such a partnership, but the truly tragic outcome is that the U.S. has withdrawn. That means that the U.S. can continue to contaminate the atmosphere, warm the earth, and pollute the oceans.

NOTES

1   National Aeronautics and Space Administration: Global Climate Change. "Scientific consensus." Available at: https://climate.nasa.gov/scientific-consensus/

2   Bulletin of the Atomic Scientists. "It is 2 minutes to midnight." Available at: https://thebulletin.org/sites/default/files/2018%20Doomsday%20Clock%20Statement.pdf

3   Climate Home News. "Leaked draft summary of UN special report on 1.5 climate goal." February 13, 2018. Available at: www.climatechangenews.com/2018/02/13/leaked-draft-summary-un-special-report-1-5c-climate-goal-full/

4   International Criminal Court. Available at: www.icc-cpi.int/Pages/Main.aspx

5   John Bacon, "China unveils market-based plan to curb global warming." *USA Today*, December 19, 2017. Available at: www.usatoday.com/story/news/world/2017/12/19/china-unveils-market-based-plan-curb-global-warming/964050001/

6   Pew Research, "What the world thinks about climate change in 7 charts." Available at: www.pewresearch.org/fact-tank/2016/04/18/what-the-world-thinks-about-climate-change-in-7-charts/. Exceptionally large number of respondents (1,000 for each country) gives us confidence in the results.

7   Available at: https://climatecommunication.yale.edu/wp-content/uploads/2017/02/Trump-Voters-and-Global-Warming.pdf

8   Available at: www.pewinternet.org/2016/10/04/the-politics-of-climate/

9   Lynn Doan and Brian Bloomberg, "Why Trump is taxing solar panels," January 23, 2018. Available at: www.washingtonpost.com/business/why-trump-is-taxing-solar-panels-imported-by-us-quicktake-qanda/2018/01/23/93cd5b34-000c-11e8-86b9-8908743c79dd_story.html?utm_term=.10de16827c68

10 National Aeronautics and Space Administration (NASA). Vital signs of the planet: Carbon dioxide. Available at: https://climate.nasa.gov/vital-signs/carbon-dioxide/

11 Union of Concerned Scientists, "Why does CO_2 get most of the attention when there are so many other heat-trapping gases?" Available at: www.ucsusa.org/global_warming/science_and_impacts/science/CO2-and-global-warming-faq.html#.WZxKmFGQyUk

12 Intergovernmental Panel on Climate Change. IPCC Fourth Assessment Report. Available at: www.ipcc.ch/publications_and_data/ar4/syr/en/figure-spm-3.html

13 Robert Glennon, "The Unfolding Tragedy of Climate Change in Bangladesh." *Scientific American*, April 21, 2017. Available at: https://blogs.scientificamerican.com/guest-blog/the-unfolding-tragedy-of-climate-change-in-bangladesh/

14 Very specifically, a persistent high-pressure ridge off the west coast of North America blocked storms from coming onshore during the winters of 2013–14 and 2014–15 and scientists associated with a particular wave pattern, which they call wavenumber-5. Haiyan Teng and Grant Branstator, "Causes of extreme ridges that induce California drought." *Journal of Climate*, February 15, 2015. DOI: 10.1175/JCLI-D-16-0524.1

15 Debra Kahn and Anne C. Mukern, "Scientists see climate change in California wildfires." *Scientific American*, October 12, 2017. Available at: www.scientificamerican.com/article/scientists-see-climate-change-in-californias-wildfires/

16 Climate Central. Available at: www.climatecentral.org/

17 Olga Khvan, "This is what Boston might look like after 2100 at varying carbon pollution levels." *Boston Magazine*, October 14, 2015. Available at: www.boston magazine.com/news/blog/2015/10/14/boston-map-climate-change/

18 Bahner Kama, "Climate migrants might reach one billion by 2050." *IPS Inter Press Service*. August 21, 2017. Available at: www.ipsnews.net/2017/08/climate-migrants-might-reach-one-billion-by-2050/

19 Jennifer Clapp, *Food* (2nd ed.). Cambridge, UK: Polity, 2016.

20 U.S. Department of Energy. "Algal biofuels." Available at: https://energy.gov/eere/bioenergy/algal-biofuels

21 Environmental Protection Agency. "Overview of greenhouse gases." Available at: www.epa.gov/ghgemissions/overview-greenhouse-gases

22 National Aeronautics and Space Administration. "Global climate change." Available at: https://climate.nasa.gov/vital-signs/carbon-dioxide/; https://climate.nasa.gov/vital-signs/global-temperature/; https://climate.nasa.gov/vital-signs/arctic-sea-ice/; https://climate.nasa.gov/vital-signs/land-ice/; https://climate.nasa.gov/vital-signs/sea-level/

23 United Nations Framework Convention on Climate Change. "Five steps to a safer future." Available at: http://unfccc.int/essential_background/convention/items/6036.php

24 The U.S. ratified this treaty; see United Nations Framework on Climate Change (Rio Declaration). Available at: http://unfccc.int/essential_background/convention/status_of_ratification/items/2631.php

25 Paris Agreement. Available at: http://unfccc.int/files/essential_background/convention/application/pdf/english_paris_agreement.pdf; www.unep.org/documents.multilingual/default.asp?documentid=78&articleid=1163

26 United Nations. Framework Convention on Climate Change: Conference of the Parties (COP). Available at: http://unfccc.int/2860.php

27 Carbon Brief. Tracking country climate pledges. Available at: www.carbonbrief.org/paris-2015-tracking-country-climate-pledges; United Nations. Framework Convention on Climate Change. NDC Registry. Available at: www4.unfccc.int/ndcregistry/Pages/All.aspx

28 In June 2017, Trump ended the U.S.'s participation in the Green Climate Fund, to the shock of world leaders and scientists.

29 UN Office of High Commissioner of Human Rights. UN expert on extreme poverty and human rights to visit the United States. Available at: www.ohchr.org/_layouts/15/WopiFrame.aspx?sourcedoc=/Documents/Issues/Poverty/Information NoteUSMission.docx&action=default&DefaultItemOpen=1

30 Carbon Brief. Seven charts show how renewable investment broke records in 2015. Available at: www.carbonbrief.org/seven-charts-show-how-renewable-investment-broke-records-in-2015

31 *The Independent*, "Solar and wind power cheaper than fossil fuels for the first time." January 4, 2017. Available at: www.independent.co.uk/environment/solar-and-wind-power-cheaper-than-fossil-fuels-for-the-first-time-a7509251.html

CHAPTER 2

Why Paris is Very Important

Initially anchoring this chapter in a review of the 2015 Paris (COP21) Agreement is helpful because this serves as a reminder that there is near universal consensus that the earth is heating up and that it is possible to slow this down if all countries cooperate. On June 1, 2017, the President of the United States announced that the U.S. would pull out of the Paris Agreement, making the U.S. the only country that is not party to the Agreement. While a formal exit from the Paris Agreement can only occur after November 4, 2020, Trump has already rolled back important domestic programs designed to slow climate change.[1] Besides, Scott Pruitt, Head of the Environmental Protection Agency (EPA), has forbidden agency employees from attending or speaking at scientific meetings devoted to climate change, and ordered the removal of all references to climate change on the EPA web page,[2] and scientists have been told that climate change research would not be funded.[3] The government continues to revoke laws that protect people's health and harm the environment, such as the law that required oil and gas companies from reporting methane emissions. Do note, too, that such actions as these accelerate the heating of the atmosphere and harm the health of everyone on the planet.

Surely, the U.S. will rejoin the Paris Agreement and the international effort to slow climate change, but it will need to play catch-up. It is important to remember that the U.S. is the world's second greatest emitter of greenhouse gases affecting the entire world. One thing, however, is fortunately the case—that American cities, businesses and households are investing in renewable energy at an ever-faster clip. Renewable energy accounts for nearly 15% of the electricity energy produced in the U.S. In fact, there are several U.S. towns that rely solely on renewable energy.

These are not entirely local responsibilities, or even national ones. Coordinated international action is urgent. A main reason is that the world's oceans cannot absorb much more CO_2 and when the tipping point is reached CO_2 will rapidly accumulate in the atmosphere, which is globally shared. Just to illustrate some of the consequences: some countries will be too hot to be habitable; the Arctic ice will rapidly melt at a faster rate, reducing the capacity of the Arctic to reflect sunlight back into space, thereby further accelerating the heating up of the planet; food, health, habitation, etc. will all be affected.

Throughout these chapters I provide evidence that unequivocally supports the full implementation of the Paris Agreement and countries meeting their targets. This requires collaboration on an unprecedented scale.[4] The majority of countries have submitted plans—Nationally Determined Contributions—indicating how they will implement it.[5] Box 2.1 includes some key provisions of the Paris Agreement. I have renumbered the provisions here for clarity since I do not include all of them. Note especially: references to [1] food security; [2] biodiversity and Mother Earth; [3] the aim for no more than 1.5°C; [4] and [5] nationally determined contribution, or NDCs; [6] non-market approaches; [7] participatory, democratic, and inclusive approaches; [8] the responsibility of developed countries to assist developing ones; and [10] provisions for withdrawing. It should be noted that the NDCs are especially important. They are the commitments made by each state as targets for implementation.[6]

Note especially that food security has priority; the emphasis on holistic agriculture (e.g., Mother Earth and biodiversity, while implicitly downplaying industrial agriculture); the intention to keep the 1.5°C limit; the importance of "nationally determined contributions" (NDCs); a comprehensive and non-market approach; and developed countries assisting developing countries (i.e., through the "Green Climate Fund").

It is important to note that cooperation and collaboration that are essential and required to prevent global catastrophe are incompatible with intense individualism and a combative, competitive attitude about rights, possessions, and privileges. Only through cooperation can we stabilize the globe.

THE SIGNIFICANCE OF THE PARIS AGREEMENT

When the gavel fell at 7:16 pm local time on Saturday, December 12, 2015 at the Paris-Le Bourget, there were shouts of joy, dancing and crying, and people were hugging and kissing each other throughout the big hall. The Paris Agreement was approved by consensus. It was subsequently opened for signature and ratification on World Earth Day,

BOX 2.1 KEY PROVISIONS OF THE PARIS AGREEMENT (COP21)[7]

[1] Recognizing the fundamental priority of safeguarding food security and ending hunger, and the particular vulnerabilities of food production systems to the adverse impacts of climate change.

[2] Noting the importance of ensuring the integrity of all ecosystems, including oceans, and the protection of biodiversity, recognized by some cultures as Mother Earth, and noting the importance for some of the concept of "climate justice," when taking action to address climate change.

[3] Holding the increase in the global average temperature to well below 2°C above pre-industrial levels and pursuing efforts to limit the temperature increase to 1.5°C above pre-industrial levels, recognizing that this would significantly reduce the risks and impacts of climate change.

[4] Each Party shall prepare, communicate and maintain successive nationally determined contributions [NDCs] that it intends to achieve. Parties shall pursue domestic mitigation measures, with the aim of achieving the objectives of such contributions.

[5] Each Party's successive nationally determined contribution will represent a progression beyond the Party's then current nationally determined contribution and reflect its highest possible ambition, reflecting its common but differentiated responsibilities and respective capabilities, in the light of different national circumstances.

[6] Parties recognize the importance of integrated, holistic and balanced non-market approaches being available to assist in the implementation of their nationally determined contributions, in the context of sustainable development and poverty eradication, in a coordinated and effective manner, including through, inter alia, mitigation, adaptation, finance, technology transfer and capacity-building, as appropriate.

[7] Parties acknowledge that adaptation action should follow a country-driven, gender-responsive, participatory and fully transparent approach, taking into consideration vulnerable groups, communities and ecosystems, and should be based on and guided by the best available science and, as appropriate, traditional knowledge, knowledge of indigenous peoples and local knowledge systems, with a view to integrating adaptation into relevant socioeconomic and environmental policies and actions, where appropriate.

[8] Developed country Parties shall provide financial resources to assist developing country Parties with respect to both mitigation and adaptation in continuation of their existing obligations under the Convention.

[9] Capacity-building under this Agreement should enhance the capacity and ability of developing country Parties, in particular countries with the least capacity, such as the least developed countries, and those that are particularly vulnerable to the adverse effects of climate change, such as small island developing states, to take effective climate change action, including, inter alia, to implement adaptation and mitigation actions.

[10] At any time after three years from the date on which this Agreement has entered into force for a Party, that Party may withdraw from this Agreement by giving written notification to the Depositary.

April 22, 2016. Months later, on Saturday, September 3, 2016, President Barack Obama and Xi Jinping formally agreed to adopt the treaty, further accelerating the clock for when the Agreement would come into force. The formal stipulation was that it would enter into force on the thirtieth day after the date on which at least 55 parties to the Convention that accounted for an estimated 55% of the total global greenhouse gas emissions ratified it. This stipulation was met on October 5, 2016, when a total of 55 parties had ratified it.[8]

In a stunning "encore" to October 5, on the very next day, October 6, the United Nations aviation agency—the International Civil Aviation Organization (ICAO)—imposed restrictions on airplane emissions, requiring airlines to buy carbon credits from designated environmental projects around the world to offset growth in emissions.[9] Also amazing, on October 15, 2016, in Kigali, Rwanda, 197 countries approved an important amendment to the Montreal Protocol to reduce and phase down hydrofluorocarbons (HFCs). In its announcement, the UN described HFCs in these terms:

Commonly used in refrigeration and air conditioning as substitutes for ozone-depleting substances, HFCs are currently the world's fastest growing greenhouse gases, their emissions increasing by up to 10 per cent each year. They are also one of the most powerful, trapping thousands of times more heat in the Earth's atmosphere than carbon dioxide (CO_2).[10]

On the last day of the Paris conference, on December 12, it appeared that the delegates would accept 2°C as the acceptable limit, but in the last hours representatives from Small Island States made persuasive enough claims for the language to read: "pursuing efforts . . . to limit the temperature increase to 1.5°C above pre-industrial levels, recognizing that this would significantly reduce the risks and impacts of climate change." (Since the Paris meeting the focus has been on 1.5°, not 2°.)

On the final evening, delegates spoke about "the world's environmental revolution." And so it seemed to be like a revolution (watching and listening from America). In fact, many countries had been on board for quite a while. Many—rather, most—have ratified other environmental treaties, and by 2015 a few countries (such as Costa Rica) could boast they depend almost entirely on renewable energy sources, placing priority on a sustainable environment for the welfare of the inhabitants.

The Paris Agreement is not a perfect agreement. It would have been better if it had specified that 1.5°C were the mandated limit; it would have been better if rich countries were legally required to pay into the Green Climate Fund to support developing countries; it would have been better had indigenous peoples and the residents of Small Island States had guarantees of assistance; and it would have been better had it included provisions for climate refugees. Nevertheless, it is an incredible achievement. The Paris Agreement is a treaty or a formal, legal commitment among parties stimulating that they will cooperatively participate in policies, actions, and agreements that will slow climate warming, and they will craft their own national practices and laws that promote this effort. It is, in short, a formal agreement based on an understanding of the basic principles of the science of climate change, an understanding of the consequences of acting or not acting, as well as a commitment to the welfare of the world's population. The future of the planet hangs in the balance, and America threatens to accelerate the planet's destruction.

Already several islands that make up the Solomon Islands have been swallowed by the sea.[11] America has not been spared. Louisiana is losing 75 square kilometers of coastal terrain every year, and the residents of Isle de Jean Charles are the first American climate refugees. Gulf Coast states are also losing land to the seas at a rapid rate.[12] There are two reasons why the oceans are rising: first, ice in both the Arctic and Antarctic is melting; second, water (the sea) expands as the air gets warmer.

It is important to highlight how quickly some countries have moved to implement plans to switch from fossil fuels to renewable energy. Scotland set an original goal of cutting emissions by 42% by 2020, but it met that goal already in 2016. In 2017, it set a new target to cut emissions by 66% within 15 years.[13] Denmark also plans to reduce its domestic greenhouse gas emissions by 40% by 2020.[14] Yet Norway has set the most ambitious emissions reduction target, committing itself to becoming carbon neutral by 2030.[15] In January 2017, China announced that it intends to spend more than $360 billion on renewable power sources like solar and wind, and it is already on track to peak its carbon dioxide emissions between 2025 and 2030.[16] Various countries—Norway, France, the Netherlands, India and the UK—have announced deadlines for when electric cars will replace internal-combustion cars.

EL NIÑO AND LA NIÑA

Before describing some of the consequences of planetary warming, it is helpful to describe the complicit roles of El Niño and La Niña. They are opposite phases of what is known as the El Niño-Southern Oscillation (ENSO) cycle—namely, fluctuating ocean temperatures in the equatorial Pacific. The warmer waters essentially slosh, or oscillate, back and forth across the Pacific, much like water in a bath-tub, causing variations in regional climate patterns. The pattern generally fluctuates between two states: warmer than normal central and eastern equatorial Pacific sea surface temperatures (El Niño), and cooler than normal central and eastern equatorial Pacific sea surface temperatures (La Niña). Many climate scientists believe that El Niño was responsible for 8–10% of the warming in 2015.[17] A Zika virus outbreak in Brazil in 2016 accompanied unusual flooding that was caused by El Niño, which shows how climate change affects human health.

ALREADY IN PERIL: CORAL REEFS AND COASTAL COMMUNITIES

Coral reefs are home to 25% of marine life and they are suffering devastating decline, especially the Great Barrier Reef off the coast of Australia. They are dying fast, and there are two causes. One is that the seas are getting warmer.[18] When coral is submerged in water that is too warm for too long, the coral polyps get stressed and spit out the algae that live inside them. Without the colorful algae, the coral flesh becomes transparent, revealing the stark white skeleton beneath, and because the algae provide the coral with 90% of its energy, the coral begins to starve.

The second cause is that when the ocean absorbs CO_2, it reduces the sea's pH or acidity level. These chemical reactions are termed "ocean acidification" or "OA" for short. The process starts when the water is too warm and the pH declines, affecting the amount of calcium carbonate in the water—the same mineral that makes up coral skeletons. As this saturation drops, corals have a harder time calcifying and any exposed skeleton dissolves.[19] Scientists believe that the oceans currently absorb

30–50% of the CO_2 produced by the burning of fossil fuels. If they did not soak up any CO_2, atmospheric CO_2 levels would be much higher than the current level of 407 parts per million. Yet it has been found that the oceans are absorbing less and less CO_2, and may reach a limit, in which case more and more would end up in the atmosphere, which would have catastrophic effects.[20] It should also be pointed out in this context that historically snow is important as it helps control how much of the sun's energy the earth absorbs. Light-colored snow and ice reflect this energy back into space, helping to keep the planet cool. However, as the snow and ice melt, it is replaced by dark land and ocean, both of which absorb energy. The amount of snow and ice loss in the last 30 years is greater than many scientists predicted, which means the earth is absorbing more solar energy than had been projected, another indication of planetary warming.[21]

Again, to stress that while acidification destroys coral reefs and shellfish, the oceans have a great capacity to absorb CO_2. If this were not the case, the earth's atmosphere, and our air, as well as the oceans, would be much warmer than they are.[22] But in fact the oceans may not have the capacity to hold so much heat for much longer and if—when—the oceans reached their saturation point, the temperature of the air would—will—accelerate very rapidly.[23]

There is more to the ocean story than the oceans' warming and acidification. Roughly 70% of the world is covered by oceans, and oceans evaporate more water as the air near the surface gets warmer. The result? More floods, more hurricanes, and more extreme precipitation events.[24] However, as the oceans' surface temperature continues to increase over time, more and more heat is released into the atmosphere. This additional heat can lead to stronger and more frequent storms, including tropical cyclones and hurricanes. Two major causes of global sea-level rise are thermal expansion caused by the warming of the ocean (since water expands as it warms) and increased melting of land-based ice, such as glaciers and ice sheets. The oceans are absorbing more than 90% of the increased atmospheric heat associated with emissions from human activity and, as a result, the oceans are warming 13% faster than earlier expected. Even worse, this warming is accelerating.[25]

Sea-level rise is also of great concern. As the sea warms, the waters expand and sea level slowly rises. Yet that is only half the story since with atmospheric and ocean warming the Greenland icecap and Antarctica are melting faster than ever anticipated, and now scientists are predicting a rise of 6 feet by the end of the century.[26] To give another estimate, over the past century, the Global Mean Sea Level (GMSL) has risen by 4–8 inches (10–20 centimeters). However, the annual rate of rise over the past

20 years has been 0.13 inches (3.2 millimeters) a year, roughly twice the average rate of the preceding 80 years.[27] All the world's cities situated on an ocean are at risk of being under water by 2100. These include New York, Miami, New Orleans, Boston, Rotterdam, Venice, Mumbai, Singapore, Osaka, Tampa, Dhaka, and Tokyo. To be sure, non-coastal communities are also at risk from storms and hurricanes, which are intensified due to climate warming. Such was the case in Texas when Hurricane Harvey made landfall in August 2017.

TRUMP PUT THE WORLD IN PERIL: WHY DON'T AMERICANS PROTEST?

Scientists from around the world were outraged when Trump withdrew the United States from the Paris Agreement. Here are a few reactions that the *Nature* journal recorded, although we have heard less from few American scientists:[28]

> President Trump's decision to introduce a request to leave the Paris agreement in 2020 is regrettable. It negates both the results of (1) serious scientific analyses (many made by US scientists) about the urgency to address the climate change problem; and (2) the rigorous assessment made by the IPCC about the technical and socio-economic aspects of response options, including their significant co-benefits in other areas like air quality, energy security, health or job creation.
>
> Jean-Pascal van Ypersele, climate scientist, Catholic University of Louvain, Louvain-la-Neuve, Belgium, and former vice-chair of the Intergovernmental Panel on Climate Change (IPCC)

> The US withdrawal from the Paris climate agreement is very disappointing and unfavourable for the United States and the rest of the world. Many climate scientists consider the Paris agreement insufficient for limiting warming to 2°C, so the task will be all the harder now. However, international climate agreements have not been very effective so far in reducing emissions, so there is still hope that the United States will proceed on other fronts, such as through bilateral agreements, clean-tech development and investing in new 'negative emissions' technologies.
>
> Atte Korhola, climate-policy and environmental-change researcher at the University of Helsinki, Finland

The US pullout is bad news for the international climate process. The United Nations negotiations need to focus on implementation. This will become more difficult, also because it is unclear how Trump wants to renegotiate the agreement. Political attention is absorbed due to the US move, attention that is needed for much more important issues such as bringing climate action forward.

Susanne Dröge, climate-policy researcher, German Institute for International and Security Affairs

Trump's decision to ignore scientific facts of climate disruption and the high risks of climate-change impacts is irresponsible not only towards his own people but to all people and life on this planet. The US administration prefers old technology over innovation and transformation. It is rejecting the enormous benefits and returns that leadership in the next industrial revolution—decarbonization—has to offer.

Thomas Stocker, former co-chair of climate science for the IPCC, and climate and environmental physicist at the University of Bern, Switzerland

Also, it should be noted that although European leaders, including Angela Merkel, Emmanuel Macron, Paolo Gentiloni, and Narendra Modi loudly protested Trump's decisions, Americans did not. Of course, there were some notable exceptions, especially Jerry Brown, Governor of California, but, by and large, Americans did not protest. The studies by Pew Research reported in the last chapter clarifies why this was the case.

IN SUMMARY

This chapter focuses mainly on the great significance of the Paris Agreement, the commitments of countries to reduce emissions, and the extraordinary urgency with which all country leaders have dedicated themselves to reducing dependency on fossil fuels. The international meetings continue, with the Poland Climate Conference (COP24) in December 2018. Annual COPs are important for many reasons, with the main one being that countries should clarify their Intended Nationally Determined Contributions (INDCs) and Nationally Determined Contributions (NDCs), and renew their commitments under the Paris Agreement. These meetings are also an opportunity for country representatives and scientists to share information about experiences about mitigation, technologies, and financing.

I stress again that the U.S. currently is a top emitter of carbon dioxide and without being a party to the Paris Agreement, it will continue to spew out carbon dioxide and thereby accelerate warming of the entire planet.

NOTES

1   *New York Times*, "Fifty-two environmental rules on their way out under Trump." October 6, 2017. Available at: www.nytimes.com/interactive/2017/10/05/climate/trump-environment-rules-reversed.html?_r=0

2   Available at: http://thehill.com/policy/energy-environment/356409-epa-removes-more-references-to-climate-change-from-its-website; *New York* magazine: "EPA prevents scientists from giving climate change talks." Available at: http://nymag.com/daily/intelligencer/2017/10/epa-prevents-scientists-from-giving-climate-change-talks.html

3   Available at: https://insideclimatenews.org/news/25082017/energy-dept-asked-scientist-remove-climate-change-project-description

4   Paris Agreement: Status of Ratification. Available at: http://unfccc.int/paris_agreement/items/9444.php

5   Interim NDC registry. Available at: www4.unfccc.int/ndcregistry/Pages/Home.aspx

6   Ibid.

7   Paris Agreement. Available at: http://unfccc.int/files/essential_background/convention/application/pdf/english_paris_agreement.pdf

8   United Nations Framework Convention on Climate Change. Paris Agreement. Available at: http://unfccc.int/paris_agreement/items/9485.php

9   Reuters, "U.N. sets limits on global airline emissions amid dissent." October 6, 2016. Available at: www.reuters.com/article/us-climatechange-aviation-idUSKCN1261QR

10 United Nations Environment Programme, "Countries agree to curb powerful greenhouse gases in largest climate breakthrough since Paris." October 15, 2016. Available at: http://unep.org/newscentre/Default.aspx?DocumentID=27086&ArticleID=36283&l=en

11 Simon Albert, Javier X. Leon, Alistair Grinham, John A. Church, Badin R. Gibbes, and Colin W.D. Woolroffe, "Interactions between sea-level rise and wave exposure on reef island dynamics in the Solomon Islands." *Environmental Research Letters*, 11 (5), May 6, 2016. Available at: http://iopscience.iop.org/article/10.1088/1748-9326/11/5/054011

12 Thomas Beller, "The residents of this Louisiana island are America's first "climate refugees." *Smithsonian*, June 29, 2016. Available at: www.smithsonianmag.com/science-nature/residents-louisiana-island-americas-first-climate-refugees-180959585/

13 Severin Carrell, "Scotland sets ambitious goal of 66% emissions cut within 15 years." January 19, 2017. Available at: www.theguardian.com/environment/2017/jan/19/scotland-sets-ambitious-goal-of-66-emissions-cut-within-15-years

14 Laurie Guevara-Stone, "Denmark reaching its goal to be fossil free through renewable energy." *The Ecologist*, March 5, 2016. Available at: www.theecologist.org/News/news_analysis/2987345/denmark_reaching_its_goal_to_be_fossil_fuel_free_through_renewable_energy.html

15   Arthur Neslen, "Norway pledges to become climate neutral by 2030." June 15, 2016. Available at: www.theguardian.com/environment/2016/jun/15/norway-pledges-to-become-climate-neutral-by-2030

16   Michael Forsythe, "China aims to spend at least $360 billion on renewable energy by 2020." January 5, 2017. Available at: www.nytimes.com/2017/01/05/world/asia/china-renewable-energy-investment.html; also see Tom Phillips, "China builds world's biggest farm in journey to become green superpower." January 19, 2017. Available at: www.theguardian.com/environment/2017/jan/19/china-builds-worlds-biggest-solar-farm-in-journey-to-become-green-superpower; Climate action tracker: China. Available at: http://climateactiontracker.org/countries/china.html

17   State Climate Office of North Carolina. Global Patterns – El Niño – Southern Oscillation (ENSO). Available at: http://climate.ncsu.edu/climate/patterns/ENSO.html. Earth Institute, Columbia University. State of the planet: El Niño and global warming – What's the connection? February 2, 2016. Available at: http://blogs.ei.columbia.edu/2016/02/02/el-nino-and-global-warming-whats-the-connection/

18   Mark D. Spalding and Barbara E. Brown, "Warm-water coral reefs and climate change." *Sciencemag.org*, November 13, 2015. Available at: www.oceanfdn.org/sites/default/files/Science-2015-Spalding-769–71.compressed.pdf

19   Marine Biodiversity and Biotechnology. "Coral pH regulation." Available at: www.cmbb.hw.ac.uk/research/coral-ecosystems/9-research/254-coral-ph-regulation.html

20   Bob Monroe, "How much CO_2 can the oceans take up? Scripps. The Keeling Curve". Available at: https://scripps.ucsd.edu/programs/keelingcurve/2013/07/03/how-much-co2-can-the-oceans-take-up/

21   See EPOCA (European Project on Ocean Acidification). "What is ocean acidification?" Available at: www.epoca-project.eu/index.php/what-is-ocean-acidification.html; National Oceanic and Atmospheric Administration. "What is ocean acidification?" Available at: http://oceanservice.noaa.gov/facts/acidification.html. Climate Reality Project. "Indicators that show climate change". Available at: www.climaterealityproject.org/blog/10-indicators-that-show-climate-change

22   C. Heinze, S. Meyer, N. Goris, L. Anderson, R. Steinfeldt, N. Chang, C. Le Quère, and D.C.E. Bakker, "The ocean carbon sink – impacts, vulnerabilities and challenges." *Earth System Dynamics*, 6, 327–357, 2015. Available at: www.earth-syst-dynam.net/6/327/2015/esd-6-327-2015.pdf

23   Cheryl Katz, "How long can oceans continue to absorb earth's excess heat?" *Yale Environment 360*, March 30, 2015. Available at: http://e360.yale.edu/features/how_long_can_oceans_continue_to_absorb_earths_excess_heat

24   National Aeronautics and Space Administration (NASA), "Storms are getting stronger". Available at: https://earthobservatory.nasa.gov/Features/ClimateStorms/page2.php

25   Lijing Cheng, Kevin E. Trenberth, John Fasullo, Tim Boyer, John Abraham, and Jiang Zhu, "Improved estimates of ocean heat content from 1960 to 2015." *Science Advances*, 3(3), March 10, 2017: e1601545; DOI: 10.1126/sciadv.1601545. Available at: http://advances.sciencemag.org/content/3/3/e1601545

26   Nicola Jones, "Abrupt sea level looms as increasingly realistic threat." May 5, 2016. Available at: http://e360.yale.edu/features/abrupt_sea_level_rise_realistic_greenland_antarctica. National Oceanic and Atmospheric Administration. "Is sea level rising?" Available at: http://oceanservice.noaa.gov/facts/sealevel.html

27 *National Geographic*. "Sea level rise." Available at: www.nationalgeographic.com/
    environment/global-warming/sea-level-rise/
28 Jeff Tollefson and Quirin Schiermeier, "How scientists reacted to the US leaving
    the US climate agreement," *Nature*, June 3, 2017. Available at: www.nature.com/
    news/how-scientists-reacted-to-the-us-leaving-the-paris-climate-agreement-1.
    22098

From the Millennium Development Goals to the Sustainable Development Goals to COP21 and the Paris Agreement

It seems quite remarkable that every country in the world—except the U.S.—would team up to approve a treaty to limit the warming of the planet and end reliance on fossil fuels by 2050, which means countries that otherwise do not respect international norms (such as North Korea and Eritrea) are party to the Paris Agreement. It seems quite amazing that such international solidarity exists that the countries that are destructive and that have few, if any, allies agree to cooperate to slow climate warming. Yes, it is in the self-interest of each country to cooperate, but in the context of the United Nations there is more than self-interest at work. The UN, from its founding in response to the defeat of the Nazis and the end of death camps, has created a world community based on trust, tolerance, and, yes, solidarity.

COP21 was an amazing achievement and based on the participation of all countries—even North Korea and Eritrea—except for one. Yet we should understand that there were precedents for such remarkable participation—notably, the adoption of the Universal Declaration of Human Rights, discussed in the next chapter, and the Millennium Development Goals (MDGs) and the Sustainable Development Goals (SDGs), discussed in this chapter. No doubt were it not for the processes that accompanied the MDGs and the SDGs, the 2015 Paris Agreement would not have been possible. By 2015, countries trusted one another.

Let's take them each in turn, starting with the MDGs (beginning in 2000 through 2015, when they were superseded by the SDGs in 2016).

The MDGs encompassed eight goals that focused entirely on the transformation of developing countries. Box 3.1 contains a list of these goals and under each there is a brief description of some of the achievements that were made by 2015, drawing from a UN report.[1] Note the importance of collaboration that underlies the goals and achievements; the Paris Agreement builds on such collaboration.

In the Preface to the 2015 final report, Ban Ki-Moon, Secretary-General of the United Nations, described the progress made toward achieving the Millennium Goals:

> the most successful anti-poverty movement in history. . . an inspiring framework . . . has enabled people across the world to improve their lives and their future prospects. The MDGs helped to lift more than one billion people out of extreme poverty, to make inroads against hunger, to enable more girls to attend school than ever before and to protect our planet. They generated new and innovative partnerships, galvanized public opinion and showed the immense value of setting ambitious goals . . . the MDGs reshaped decision-making in developed and developing countries alike.[2]

What, of course, made all of this possible is that each of 189 countries volunteered to be a part of this process, whether as a developed or developing country. It was a thoroughly collaborative and cooperative process, setting the precedence for the SDGs and the Paris Agreement. Yet there were no goals in the MDGs aimed at reducing greenhouse gases or combating climate change although throughout this period, from 2000 to 2015, there already were annual international UN-sponsored conferences devoted to climate change. In fact, as earlier noted, there had been world conferences since 1979 devoted to understanding and measuring the extent to which greenhouse gases are warming the earth.

However, in the final MDG 2015 report attention focused on climate change, providing a launch for the 2016 Sustainable Development Goals (SDGs), which more explicitly address climate change. This 2015 report states:

> Climate change and environmental degradation undermine progress achieved, and poor people suffer the most. Global emissions of carbon dioxide have increased by over 50 per cent since 1990. Addressing the unabated rise in greenhouse gas emissions and the resulting likely impacts of climate change, such as altered ecosystems, weather extremes and risks to society, remains an urgent, critical challenge for the global community.[3]

BOX 3.1 MILLENNIUM GOALS, 2000–2015: ACHIEVEMENTS BY 2015

Goal 1: Eradicate Extreme Poverty and Hunger

- In 1990, nearly half of the population in the developing world lived on less than $1.25 a day. That dropped to 14% of the population in 2015.
- Globally, the number of people living in extreme poverty has declined by more than half, falling from 19 billion in 1990 to 836 million in 2015.
- The proportion of undernourished people in developing countries has fallen by almost half since 1990, from 23.3% in 1990–1992 to 12.9% in 2014–2016.

Goal 2: Achieve Universal Primary Education

- The primary school net enrollment rate in the developing regions reached 91% in 2015, up from 83% in 2000.
- The number of out-of-school of primary school age worldwide has fallen by almost half, to an estimated 57 million in 2015, down from 100 million in 2000.
- The literacy rate among youth aged 15 to 24 has increased globally from 83% to 91% between 1990 and 2015.

Goal 3: Promote Gender Equality and Empower Women

- The developing regions as a whole have achieved the target to eliminate gender disparity in primary, secondary and tertiary education.
- In Southern Asia, only 74 girls were enrolled in primary school for every 100 boys in 1990. Today, 103 girls are enrolled for every 100 boys.
- Women now make up 41% of paid workers outside the agricultural sector, an increase from 35% in 1990.

Goal 4: Reduce Child Mortality

- The global under-five mortality rate has declined by more than half, dropping from 90 to 43 deaths per 1,000 live births between 1990 and 2015.
- Since the early 1990s, the rate of reduction of under-five mortality was over five times faster during 2005–2013 than it was during 1990–1995.
- About 84% of children worldwide received at least one dose of measles-containing vaccine in 2013, up from 73% in 2000.

Goal 5: Improve Maternal Health

- Since 1990, the maternal mortality rate has declined by 45% worldwide.
- More than 71% of births were assisted by skilled personnel globally in 2014, an increase from 59% in 1990.
- Contraceptive prevalence among women, aged 15 to 49, married or in a union, increased from 55% worldwide to 64% in 2015.

Goal 6: Combat HIV/AIDS, Malaria, and Other Diseases

- By June 2014, 13.6 million people living with HIV were receiving antiretroviral therapy (ART) globally, an immense increase from just 800,000 in 2003. ART averted 7.6 million deaths from AIDS between 1995 and 2013.
- Over 6.2 million malaria deaths were averted between 2000 and 2015. The global malaria incidence rate has fallen by an estimated 37% and the mortality rate by 58%.
- Between 2000 and 2013, tuberculosis prevention, diagnosis and treatment interventions saved an estimated 57 million lives.

Goal 7: Ensure Environmental Sustainability

- Ozone-depleting substances have been virtually eliminated since 1990 and the ozone layer is expected to recover by mid-century.
- In 2015, 91% of the global population is using an improved drinking-water source, compared to 76% in 1990.
- Worldwide, 2.1 billion people have gained access to improved sanitation. The proportion of people practicing open defecation has fallen almost by half since 1990.

Goal 8 Develop a Global Partnership for Development

- Official development assistance from developed to developing countries increased by 66% from 2000 to 2014 to reach $135.2 billion.
- The number of mobile-cellular subscriptions grew almost tenfold in 15 years, from 738 million in 2000 to over 7 billion in 2015.
- Internet penetration grew from just over 6% of the world's population in 2000 to 43% in 2015. As a result, 3.2 billion people are linked to a global network of content and applications.

THE SDGs

The context is extremely important. Keep in mind that the focus of the MDGs was on developing countries, but accompanying the expansion of neoliberalism and globalization has been increasing economic inequality, so while the world proclaimed that rich countries should help poor countries, inequality has actually been increasing – that is, people in poor countries have been getting poorer and people in rich countries have been getting richer.[4] Not only that, rich countries have been throwing carbon dioxide up into the atmosphere for several centuries, as noted earlier, and now they ask poor countries not to industrialize in the same way, but instead to adopt and use renewable energy. Wind turbines and solar panels are, of course, expensive.

Recall that in Durban in 2011, at COP17, the parties agreed to establish the Green Climate Fund that would assist poor countries to invest in renewable energy and to develop resilient strategies in dealing with climate change. At COP18, which met in Doha in 2012, decisions were reached about compensation to poor countries that experienced grave harms from climatic events. The SDG Summit met in New York City and on September 25, 2015, the goals were adopted by all the world's nations to "cover nearly every aspect of our future—for our planet, and for humankind." Besides, it was said, "they concern all people, all countries, and all parts of society."[5]

The context here is important. First, and to repeat, the SDGs apply to all countries, not just to developing countries. Second, the SDGs reflect the understanding that the global community must tackle and reduce inequalities. Third, it was recognized that sustainable development is only possible if the pace of climate warming is slowed and if all countries commit to ensuring that carbon dioxide emissions are reduced to zero by 2040–2050. Only this would make it likely that global warming is reduced to 1.5°C. Box 3.2 lists the SDGs.

Each of the 17 goals has specific targets to be achieved by 2030, with a total of 169 targets. These can be viewed online,[6] but our attention here focuses on Goal 10—reduce inequality—and Goal 13—combat climate change. Let's take the target of Goal 10 first, noting that the reduction of inequality has broad and far-reaching implications for other goals. Reducing inequality also addresses Goal 1 (ending poverty), Goal 2 (ending hunger, achieving food security, and improving nutrition), and Goal 8 (promoting sustained, inclusive, and sustainable economic growth). Inequality abounds in the contemporary world; it is worse than it has been in decades and it steadily grows and grows.[7]

BOX 3.2 SUSTAINABLE DEVELOPMENT GOALS, 2015–2030[8]

1. End poverty in all its forms everywhere.
2. End hunger, achieve food security and improved nutrition, and promote sustainable agriculture.
3. Ensure healthy lives and promote well-being for all at all ages.
4. Ensure inclusive and equitable quality education and promote lifelong learning opportunities for all.
5. Achieve gender equality and empower all women and girls
6. Ensure availability and sustainable management of water and sanitation for all.
7. Ensure access to affordable, reliable, sustainable and modern energy for all.
8. Promote sustained, inclusive and sustainable economic growth, full and productive employment, and decent work for all.
9. Build resilient infrastructure, promote inclusive and sustainable industrialization, and foster innovation.
10. Reduce inequality within and among countries.
11. Make cities and human settlements inclusive, safe, resilient and sustainable.
12. Ensure sustainable consumption and production patterns.
13. Take urgent action to combat climate change and its impacts*
14. Conserve and sustainably use the oceans, seas and marine resources for sustainable development
15. Protect, restore and promote sustainable use of terrestrial ecosystems, sustainably manage forests, combat desertification and halt and reverse land degradation, and halt biodiversity loss.
16. Promote peaceful and inclusive societies for sustainable development, provide access to justice for all and build effective, accountable and inclusive institutions at all levels.
17. Strengthen the means of implementation and revitalize the global partnership for sustainable development.

* Acknowledging that the United Nations Framework Convention on Climate Change is the primary international, intergovernmental forum for negotiating the global response to climate change.

Much economic inequality—in wealth, income, or wages—is inherently unfair. This is so because people have a great variety of talents, skills, backgrounds, and interests, and these cannot be ranked or even compared. Who has the more complicated job—a farmer, a teacher, an accountant, or a plumber? We need talented and motivated farmers just as we need talented and motivated teachers, accountants, and plumbers. To be sure, being, say, a farmer, teacher, accountant or plumber in Cambodia or other developing country is probably more challenging and complicated than working in the same occupation in any developed country. Isn't it wrong that farmers, teachers, accountants, and plumbers in rich countries make so much more than their counterparts in, say, Cambodia, or anywhere in the global South? Nor can one say that there is any justification to the fact that the standard of living is higher in the U.S. and Europe compared with the rest of the world. Yet that is exactly the case. These are the considerations that inform the specific targets of Goal 10, as listed in Box 3.3.

Taken together, these ten targets, when implemented, would make an extraordinary difference, reducing the great inequalities between people living in rich countries and those living in poor countries, as well as the great inequalities within countries. Most of the targets are self-explanatory, but numbers 8, 9 and 10 need clarification. Number 8 states that developing countries have yet to receive special and differential treatment in accordance with WTO agreements. This has been a problem that has dragged on without being resolved.[9] Number 9 highlights the importance of official development assistance, or ODA, which is the percentage of gross national income that developed countries are expected to give to developing countries, and is currently set as a percentage of gross national income at 0.7%. Developed countries that meet that objective are Sweden, the United Arab Emirates, Norway, Luxembourg, Denmark, the Netherlands, and the United Kingdom.[10] And number 10 refers to the transaction costs of remittances that migrants living and working in a host country must pay. Worldwide remittances in 2015 were $592 billion (in US dollars), and the transaction costs now paid by senders average 7.4%, more than double that recommended in goal 10.[11]

Box 3.4 lists the targets for Goal 13 related to climate change. The Paris Climate Change Summit occurred on the heels of the Sustainable Development Goals Summit (30 November to 12 December, and 25–27 September, respectively), and for that reason the targets are not detailed. However, they include a focus on assistance to developing countries by clarifying the importance of the Green Climate Fund (Target 13.a) that assists developing countries acquire renewable energy technologies. Target 13.b reinforces Target 13.a and also highlights the special importance of small island states that are at peril.

BOX 3.3 SDG GOAL 10: TARGETS TO REDUCE INEQUALITY WITHIN AND AMONG COUNTRIES[12]

1. By 2030, progressively achieve and sustain income growth of the bottom 40% of the population at a rate higher than the national average.
2. By 2030, empower and promote the social, economic and political inclusion of all, irrespective of age, sex, disability, race, ethnicity, origin, religion, or economic or other status.
3. Ensure equal opportunity and reduce inequalities of outcome, including by eliminating discriminatory laws, policies and practices, and promoting appropriate legislation, policies and action in this regard.
4. Adopt policies, especially fiscal, wage and social protection policies, and progressively achieve greater equality.
5. Improve the regulation and monitoring of global financial markets and institutions, and strengthen the implementation of such regulations.
6. Ensure enhanced representation and voice for developing countries in decision-making in global international economic and financial institutions in order to deliver more effective, credible, accountable and legitimate institutions.
7. Facilitate orderly, safe, regular and responsible migration and mobility of people, including through the implementation of planned and well-managed migration policies.
8. Implement the principle of special and differential treatment for developing countries, in particular the least developed countries, in accordance with World Trade Organization agreements.
9. Encourage official development assistance and financial flows, including foreign direct investment, to states where the need is greatest, in particular least developed countries, African countries, small island developing states and landlocked developing countries, in accordance with their national plans and programmes.
10. By 2030, reduce to less than 3% the transaction costs of migrant remittances and eliminate remittance corridors with costs higher than 5%.

BOX 3.4 SDG GOAL 13: TAKE URGENT ACTION TO COMBAT CLIMATE CHANGE AND ITS IMPACT[13]

1. Strengthen resilience and adaptive capacity to climate-related hazards and natural disasters in all countries.
2. Integrate climate change measures into national policies, strategies and planning.
3. Improve education, awareness-raising, and human and institutional capacity on climate change mitigation, adaptation, impact reduction and early warning.
4. Implement the commitment undertaken by developed country parties to the United Nations Framework Convention on Climate Change to a goal of mobilizing jointly $100 billion annually by 2020 from all sources to address the needs of developing countries in the context of meaningful mitigation actions and transparency on implementation, and fully operationalize the Green Climate Fund through its capitalization as soon as possible.
5. Promote mechanisms for raising capacity for effective climate change-related planning and management in the least developed countries and developing Small Island States, including focusing on women, youth, and local and marginalized communities.

CONCLUSION: THE WORLD'S TWO BIGGEST PROBLEMS

Global inequality has never been worse. According to one estimate, the richest 1% has more wealth than the rest of the world's population combined.[14] Another way of looking at this is to ask how many people are poor. Almost half the world—over 3 billion people—live on less than $2.50 a day. At least 80% of the world's people live on less than $10 a day. One way that people from the developed world should look at climate change is to understand that for the most part the developed world caused it, while benefiting from early industrialization. This is why the Green Climate Fund was established. And yet, more generally, one must be concerned that these horrific inequalities will greatly impair the degree of global cooperation that is required to slow the heating of the planet. Hungry people do not buy solar panels.

Yet facing all the world's peoples is the most horrific and comprehensive planetary disaster imaginable. Climate change will spare none, although it cannot be said that everyone will be equally affected. The Middle East will be intolerably hot. Many, if not most, of the Small Island States will be overcome by the sea. Few coastal cities will be spared. The poor who already suffer will suffer more. To be sure, climate change not only means warming. It will also mean the exacerbation of health problems and displaced populations, hunger, flooding, and violent storms. Because Trump gave the United States the license to abandon the constraints of the Paris Agreement and to abandon cooperation with the rest of the world, the U.S. will be justifiably be held in contempt by the world's peoples.

NOTES

1   United Nations, The Millennium Goals Report 2015. Available at: www.un.org/
    millenniumgoals/2015_MDG_Report/pdf/MDG%202015%20rev%20(July%201).
    pdf
2   Ibid.
3   Ibid.
4   Lant Pritchard, "Divergence big time." Policy Research Working Papers. The
    World Bank. Available at: http://elibrary.worldbank.org/doi/abs/10.1596/1813-
    9450-1522; Sudhir and Paul Segal, "The distribution of income." Chapter 11 in
    A.B. Atkinson and F. Bourguignon, Handbook of Income Distribution, Volume 2A,
    Elsevier, Amsterdam, 2015. Available at: www.economics.ox.ac.uk/materials/
    papers/13376/anand-segal-handbook-pdf-mar15.pdf
5   Learn about the SDGs. Available at: http://17goals.org/
6   Sustainable Development Knowledge Platform. Available at: https://sustainable
    development.un.org/post2015/transformingourworld
7   Facundo Alvaredo, Lucas Chancel, Thomas Piketty, Immanuel Saez, and Gabriel
    Zucman, "Global Inequality Dynamics." National Bureau of Economics Research.
    Working Paper, No. 23119, February 2017. Available at: https://eml.berkeley.edu/
    ~saez/ACPSZ2017NBERWP.pdf; Anthony B. Atkinson, Inequality: What Can
    Be Done? Cambridge, MA: Harvard University Press, 2015; Joseph E. Stiglitz,
    The Great Divide: Unequal Societies and What We Can Do About Them. New York:
    W.W. Norton, 2015.
8   Sustainable Development Knowledge Platform. Available at: https://sustainable
    development.un.org/post2015/transformingourworld
9   Aurelie Walker, "The WTO has failed developing nations." The Guardian,
    November 14, 2011. Available at: www.theguardian.com/global-development/
    poverty-matters/2011/nov/14/wto-fails-developing-countries. Aileen Kwa, "WTO
    and Developing Countries." Foreign Policy in Focus. Available at: http://fpif.org/
    wto_and_developing_countries/
10 OECD, "Development aid in 2015 continues to grow despite costs for in donor
    refugees: 2015 preliminary ODA figures." Available at: www.oecd.org/dac/stats/
    ODA-2015-detailed-summary.pdf

11   The World Bank. "Remittances to developing countries edge up slightly in 2015."
      April 13, 2016. Available at: www.worldbank.org/en/news/press-release/2016/04/
      13/remittances-to-developing-countries-edge-up-slightly-in-2015. Caroline Freund
      and Nikola Spatafora, "Remittances, transaction costs, and informality." *Journal of
      Development Economics*, *86*(2), June 2008. Available at: www.sciencedirect.com/
      science/article/pii/S0304387807000818

12   Sustainable Development Knowledge Platform. Available at: https://sustainable
      development.un.org/post2015/transformingourworld
      The individual targets are renumbered here for clarity.

13   Sustainable Development Knowledge Platform. Available at: https://sustainable
      development.un.org/post2015/transformingourworld

14   Oxfam-Credit Suisse. "Global wealth report again exposes inequality." November
      22, 2016. Available at: www.oxfam.org/en/pressroom/reactions/credit-suisse-
      global-wealth-report-again-exposes-shocking-inequality

CHAPTER 4

What are Human Rights?

Human rights can be succinctly summarized: "Equality and dignity for everyone." In greater detail: we all are equally entitled to what makes us human and to live our lives with dignity—namely, to have and enjoy an identity, food, water, social security, religion and worship, shelter, education, healthcare and medicine, employment, non-discrimination, an adequate standard of living, a sustainable environment and protection from climate change, civil and political rights, peace, the right to live in a society that promotes economic equality and unequivocally upholds equal standards of justice for everyone, and that provides opportunities for everyone to participate in the affairs of their community.

Human rights were internationally proclaimed in the Universal Declaration of Human Rights (UDHR), an agreement that was approved by the state parties of the UN General Assembly on December 10, 1948. It is an agreement, not a treaty, but as I discuss below, it is the foundation of eight human rights treaties, each of which has been ratified by a majority of states. The implications of climate change for human rights are extraordinary. These include the provision and access to food and water, security for climate refugees, and protection from diseases caused or aggravated by climatic warming, such as malaria and dengue.

There is one thing that everybody on the planet should know and that is that observing and protecting human rights must take the highest priority in the coming decades. If that does not happen, there will be fierce and ugly competition for food, water, and all resources. The intense individualism that is at the very core of American society does not easily lend itself to the sort of cooperation that will be required.

Because food is a human right—indeed, the most essential human right—and because climate change will greatly affect the quality and quantity of food available, I devote the next two chapters to food. Keep in mind that the consequence of Trump's leaving the Paris Agreement is that warming will accelerate, with immense detrimental consequences to food crops.

THE UNIVERSAL DECLARATION OF HUMAN RIGHTS

The UDHR is remarkably clear and succinct. Specifically, it encompasses civil and political rights as laid out in the 1791 U.S. Bill of Rights and the 1789 *Déclaration des droits de l'homme et du citoyen*, but goes beyond these to define and protect economic, social and cultural rights. The Preamble states that recognition of the inherent dignity and of the equal and inalienable rights of all members of the human family is the foundation of freedom, justice and peace. Article 1 states the key principle: All human beings are born free and equal in dignity and rights. They are endowed with reason and conscience and should act toward one another in a spirit of brotherhood. Article 2 emphasizes that everyone is entitled to all the rights and freedoms set forth in this declaration, without distinction of any kind, such as race, color, sex, language, religion, political or other opinion, national or social origin, property, birth or other status, or political, jurisdictional or international status of the country or territory to which a person belongs.

Articles 3 through 20 affirm persons' political and civil rights—namely, the abstracted legal rights individuals have vis-à-vis the state. Although the U.S. Constitution was the first to elaborate political and civil rights, by now virtually all states are committed to protecting these rights of citizens. Briefly, they include the right not to be tortured, equal recognition before the law, the right to a fair trial, freedom of assembly, freedom of opinion, freedom of movement, right to a nationality, right to own property, freedom to hold opinions, and right to take part in government.

One way to distinguish these rights in Articles 3 through 20 from those spelled out in Articles 21 through 30 is to consider Articles 3 through 20 to be abstracted legal rights that chiefly protect persons against a potentially obtrusive and abusive state, while Articles 21 though 30 clarify the rights of humans as conscious, ethical, social and corporal beings. These are especially relevant when considering that humans need protection through what are likely to be horrific climatic events. Article 21 is the right to take part in one's government. Article 22 is the right to social security. Article 23 is the right to work, to join a union, and the right to equal pay for equal work. Article 24 is the right to rest and leisure. Article 25 is the right to a standard of living adequate for health, food, clothing, housing, medical care, necessary social services, security in the event of sickness or lack of livelihood. Article 26 is the right to education (free at least in the primary grades) and education that promotes understanding, tolerance and friendships among all nations, racial or religious groups, and education that provides for the advance of peace. Article 27 highlights the right to culture and to share in the advance of science. Article 28 clarifies that everybody has the right to an international order in which these rights and freedoms can be realized. Article 29 contends that everyone has duties to their community. Article 30 affirms the rights and freedoms set forth in the UDHR.

As noted, the UDHR never became a treaty, but now most states' constitutions include fundamental human rights, as well as civil and political rights. The U.S. is the exception. Its constitution includes only civil and political rights, and does not include social, economic, and cultural rights. (Yes, in the U.S. many social, economic and cultural rights are protected by law, but laws can be rescinded or ignored.) Here are some examples of constitutional rights:[1]

• Nepal: Each citizen shall have the right to food.
• Armenia: discrimination based on sex, race, skin color, ethnic or social origin, genetic features, language, religion, worldview, political or any other views, belonging to a national minority, property status, birth, disability, age, or other personal or social circumstances shall be prohibited.
• France shall be an indivisible, secular, democratic and social republic. It shall ensure the equality of all citizens before the law, without distinction of origin, race or religion. It shall respect all beliefs. It shall be organized on a decentralized basis.
• Finland: The Sami, as an indigenous people, as well as the Roma and other groups, have the right to maintain and develop their own language and culture.
• South Africa: Everyone has the right to have access to sufficient food and water.
• Belize: The right to work and the pursuit of happiness, which protect the identity, dignity and social and cultural values of Belizeans, including Belize's indigenous peoples.

When rights are protected legally—constitutionally—they are easier to defend, and when social, economic and cultural rights are protected legally, they help to unify the entire society. By this, I mean, for example, that everyone has a stake in ensuring a minimum wage or quality education.

HUMAN RIGHTS TREATIES

The UDHR was affirmed by consensus on June 25, 1993 in Vienna at the World Conference on Human Rights. It is not a treaty. However, specific components or sections of the UDHR are formalized as treaties. These are listed in Box 4.1.[2] Both the number of ratifiers is reported as well as the number of signatories—and the maximum in both cases is 197—but they are not mutually exclusive so that a state can be both a signatory and party, However, being a signatory carries little responsibility and does not require regular review. Being a party to a treaty means that the state is held accountable for upholding the treaty and for participating in review processes. The human rights treaty with the most ratifiers is the Convention on the Rights of the Child. Indeed, every single country, except the United States, has ratified this treaty. In fact, the U.S. has not ratified a single treaty, with the occasional disclaimer that the "treaty is not self-executing." In other words, the U.S. has not supported or endorsed human rights, or even cooperated with the international community to clarify, define or enforce international human rights. Nor has the U.S. included human rights—except for civil and political rights—in its Constitution.[3]

HUMAN RIGHTS AS PRACTICE AND IN LAW

It is the right of *everyone* to have food, housing, education, healthcare, and all the basic things needed to live with security as equals and with human dignity. In developed countries the trajectory is reassuring since the typical pattern is security through adolescence by virtue of family membership, security achieved in adulthood through employment or through other support as adults, and finally, security achieved through retirement programs as older adults. The majority of state constitutions—131 out of 190—explicitly refer to healthcare as a basic human right, and 83 state constitutions state that a reasonable standard of living is a fundamental right.[4]

Human rights are also incorporated into the treaties of the Organization of American States, the African Union, and the Charter of the European Union. To give an example, in one treaty of the Organization of American States—the Protocol of San Salvador—sections are devoted to the right to healthcare, education, food, satisfactory conditions of work, and protection of children, disabled persons, and the elderly. The majority of member states have ratified the treaty, but not the United States.[5] In other words, the U.S. only officially recognizes political and civil rights, and does not unconditionally recognize social, cultural, or economic rights.

BOX 4.1 HUMAN RIGHTS TREATIES, YEAR OF GENERAL ASSEMBLY APPROVAL, NUMBER OF STATE PARTIES, WHETHER THE U.S. IS A PARTY OR NOT, U.S. RESERVATIONS, NUMBER OF SIGNATORIES, WHETHER THE U.S. IS A SIGNATORY OR NOT (MAY 1, 2017)[6]

International Covenant on Economic, Social and Cultural Rights (CESCR), 1966
165 parties; U.S. is not a party.
71 signatories; U.S. is a signatory.

International Covenant on Civil and Political Rights (CCPR), 1966
169 parties; U.S. is a party; "provisions are not self-executing."
7 signatories; U.S. is a signatory.

International Convention on the Elimination of All Forms of Racial Discrimination (CERD), 1965
178 parties; U.S. is a party; "provisions are not self-executing."
88 signatories; U.S. is a signatory.

Convention on the Elimination of All Forms of Discrimination against Women (CEDAW), 1979
189 parties; U.S. is not a party.
99 signatories; U.S. is a signatory.

Convention against Torture and Other Cruel, Inhuman or Degrading Treatment or Punishment (CAT), 1984
161 parties; U.S. is a party; "provisions are not self-executing."
83 signatories; U.S. is a signatory.

Convention on the Rights of the Child (CRIC), 1989
196 parties; U.S. is not a party.
140 signatories; U.S. is a signatory.

International Convention on the Protection of the Rights of All Migrant Workers and Members of their Families (CRMW), 1990
51 parties; U.S. is not a party.
38 signatories; U.S. is not a signatory.

Convention on the Rights of Persons with Disabilities, 2006
173 parties; U.S. is not a party.
160 signatories; U.S. is a signatory.

Nevertheless, it is important to add that many, if not most, Americans do support many of the things that people in other countries call human rights, including children's rights, gender identity, gay rights, nondiscrimination, equality for women and men, education, and, perhaps, healthcare and housing. Of course, there are laws, programs and funding that pertain to all of these, but there is no principled consensus or constitutional basis for human rights in the U.S. It is important to note, however, that American laws and programs, for, say, Supplemental Nutrition Assistance Program (SNAP), which provides food assistance to low-income Americans, can be rescinded at any time, just as the Affordable Care Act has been virtually abolished by the Trump administration. Unlike most countries, the U.S. does not recognize that social and economic rights need constitutional protection. However, it is interesting to speculate that everyone has an intuitive sense of human rights, which I suspect we will need when we say that we have lost our home because it is under water.

HUMAN RIGHTS AND CLIMATE CHANGE

Climate change will accompany insufferable heat for millions of people, dead farm animals, dried and inedible crops, incessant floods, sea rise and outbreaks of deadly diseases. Indeed, climate change is already having such effects as these in many parts of the world. Islands have already disappeared. Only a sliver of the Isle de Jean Charles remains off the coast of Louisiana; islands off the Virginia and Alaska coasts will be gone within a few years; Hurricane Maria decimated Puerto Rico in September 2017; Harvey hit Houston, Texas, in August 2017. It is estimated that already there are over one million climate refugees. It is predicted that southern Spain will become a desert, and already the deserts of sub-Saharan Africa have expanded and have become hotter.

It should be noted, too, that about 26% of the Netherlands is under sea level, and 70% of the total area of Bangladesh is less than 1 meter (about 3 feet) above sea level. Both countries are extremely vulnerable as the seas rise. Americans have borne the brunt of storms that scientists attribute to climate change, notably Harvey, but also Sandy and Katrina. Coastal American cities that are likely to be completely or mostly flooded by 2100 include Boston, Miami, Fort Lauderdale, New York City, Atlantic City, Honolulu, New Orleans, San Diego, Los Angeles, Charleston, Virginia Beach, Seattle, and Savannah.[7] This means there will be unprecedented movement of people away from coastal cities to inland communities. In other words, sea rise will affect billions of Americans.

Under these conditions there must be a powerful collective commitment to protect human rights, otherwise there will be chaos.

It will be the responsibility of states to uphold human rights principles by, for example, ensuring that there is enough food and water, provisions for medicine, and adequate housing. States need to ensure accountability and universal protection, and to uphold the responsibility to protect. Because climate change creates new challenges for the way human rights must be protected, it is helpful to quote from the summary issued by the Office of the High Commissioner of Human Rights (OHCHR) following a two-day meeting on climate change and human rights:

> Critically, it is not enough to simply focus on ensuring that climate actions respect human rights. A rights-based approach requires states to take affirmative action to respect, protect, promote and fulfill all human rights for all persons. Failure to prevent foreseeable human rights harms caused by climate change, or at the very least to mobilize maximum available resources in an effort to do so, constitutes a breach of this obligation. Human rights obligations apply to the goals and commitments of states in the area of climate change and require that climate actions should focus on protecting the rights of all those vulnerable to climate change, starting with those most affected.
>
> State commitments therefore require international cooperation, including financial, technological and capacity-building support, to realize low-carbon, climate-resilient, and sustainable development, while also rapidly reducing greenhouse gas emissions. Only by integrating human rights in climate actions and policies, and empowering people to participate in policy formulation, can states promote sustainability and ensure the accountability of all duty-bearers for their actions.[8]

Note especially that this statement stresses *the role of the state in reducing or slowing climate by creating institutions and technologies that curb climate change and reduce emissions in order to uphold human rights and protect and empower people.* Certainly, this is consistent with the philosophical premises of the UDHR, and the horrific consequences of climate warming obviously strengthen the assumptions made by the human rights community. In other words, the assumption here is that not only will and must states be empowered to uphold rights such as food, water, education and so forth, they are responsible for ensuring that there are institutions and technologies in place through which basic human rights can be realized as the planet heats up.

CONCLUSION

The United States is relatively unique, first, because it does not ratify human rights treaties, and, second, because it has not revised its own constitution to include human rights. The reason in both cases may very well be that its traditions are bound up with capitalism, or we could say, every man, woman, and child for himself or herself—sink or swim; there is no free lunch. There are no basic protections or no affirmation of solidarity. This means that as the planet heats up, America could very well become a free-for-all. Worse, without being a party to the Paris Agreement, the U.S. could accelerate the heating of the planet and harm everyone on the planet.

Here, I can give an example of the way that climate change has affected millions of people already. It is well known that Syrians are risking their lives to flee Assad's brutality, but what has not been reported very much by the Western media is that there has been a severe drought that has driven Syrians off their lands to flee, risking their lives and abandoning everything that they own. Syria is losing its groundwater. It became impossible to water the crops. There was nothing to drink as the wells went dry.[9] In fact, what is happening globally—and most particularly in the Middle East—is that groundwater is declining at an alarming rate.

Climate change will create havoc with everyone's lives, and is already doing that in many parts of the world. Because the U.S. left the Paris Agreement, it will accelerate warming all over the globe. This is why I say that Trump has created a crime against humanity.

NOTES

1   Constitute Project. Available at: www.constituteproject.org/search?lang=en
2   Note: there are Optional Protocols, but they are not listed since most set up and describe the review procedures, and are not substantive.
3   See Judith Blau and Louis Edgar Esparza, *Human Rights: A Primer* (2nd ed.) New York: Routledge, 2016.
4   Constitute Project. Available at: www.constituteproject.org/
5   Additional Protocol to the American Convention on Human Rights in the Area of Economic, Social and Cultural Rights. "Protocol of San Salvador," Organization of American States. Text available at: www.oas.org/juridico/english/treaties/a-52.html; ratifications available at: www.oas.org/juridico/english/sigs/a-52.html
6   Virtually all of the optional protocols set up mechanisms for review of treaty compliance, although there are a few substantive ones, to none of which is the U.S. a party. Substantive ones are the Second Optional Protocol to the International Covenant on Civil and Political Rights aiming at abolition of the death penalty; Optional Protocol to the Convention on the Rights of the Child on the involvement

of children in armed conflict; Optional Protocol to the Convention on the Rights of the Child on the sale of children, child prostitution and child pornography; Optional Protocol to the Convention on the Rights of the Child on a communications procedure. Source: UN Office of the High Commissioner of Human Rights. Available at: www.ohchr.org/EN/ProfessionalInterest/Pages/CoreInstruments.aspx; http://treaties.un.org/Pages/Treaties.aspx?id=4&subid=A&lang=en

7   See, for example, Matthew E. Hauer, Jason M. Evans and Deepak R. Mishra, "Millions projected to be at risk from sea-level rise in the continental United States." *Nature Climate Change*, 6: 691–695, March 14, 2016. Available at: www.nature.com/nclimate/journal/v6/n7/full/nclimate2961.html

8   Office of the High Commissioner for Human Rights, "Understanding Human Rights and Climate Change." Available at: www.ohchr.org/Documents/Issues/ClimateChange/COP21.pdf

9   John Wendle, "Ominous story of Syria climate refugees." *Scientific American*, December 17, 2015. Available at: www.scientificamerican.com/article/ominous-story-of-syria-climate-refugees

CHAPTER 5

Is Food a Human Right or a Commercial Product?

Food is an unconditional right of all people. It is assumed, even proclaimed so, in religious texts, spelled out in the Quran and the Bible, especially, but also in Baha'i and Hindi texts. For example, Quran 2:168 states, "Eat of what is lawful and wholesome on the earth"; Quran 2:172 is "Eat of the good things which we have provided for you"; and Genesis 9.3 is "Every moving thing that lives shall be food for you. And as I gave you the green plants, I give you everything." The right to food is also enshrined in national laws as well as regional and international declarations and treaties. For example, the constitution of Belarus includes this provision: "Everyone has the right to a decent standard of living, including appropriate food, clothing, housing and a continuous improvement of conditions necessary to attain this." And the Fiji constitution includes this statement: "The state must take reasonable measures within its available resources to achieve the progressive realisation of the right of every person to be free from hunger, to have adequate food of acceptable quality and to clean and safe water in adequate quantities."[1]

Of course, food is a human right. We could not survive without adequate food, which is intuitively clear and comprehensively documented.[2] Americans are at risk because the right to food is not constitutionally protected in the U.S. In fact, it is estimated that one in six Americans go hungry and this includes 1.6 million children. We Americans have no fundamental right to food. Besides revising the U.S. Constitution, we must anticipate shortages of food and water as the heating of the planet accelerates.

There are two huge challenges ahead. The first is ensuring that there is no hunger anywhere and that all peoples have enough to eat, and in this chapter I provide an overview of hunger and the right to food. The second is ensuring that food is healthy, and I raise questions here about the healthiness of industrial agriculture that predominates in the U.S. I turn to a discussion of climate and agroecology in Chapter 6, contending that agroecology is superior to industrial agriculture in several respects. In this chapter, I sketch out our understanding of the principle that food is a right, and also contend that it is a positive right, and clarify how industrial agriculture has not only contributed to global warming, but has done so by disempowering farmers and risking food quality.

This chapter is based on the principle that all people deserve food. With tumultuous weather ahead, in the decades to come, this principle will ensure that we all remain fully human—which is to say, civil, compassionate—and, yes—fully deserving.

FORMALIZING THE RIGHT TO FOOD

In the American context, the right to food was first formally laid out by Franklin Delano Roosevelt (FDR), who in his State of the Union address on January 11, 1944, stated:

> In our day these economic truths have become accepted as self-evident. We have accepted, so to speak, a second Bill of Rights under which a new basis of security and prosperity can be established for all regardless of station, race, or creed.
>
> Among these are:
>
> – The right to a useful and remunerative job in the industries or shops or farms or mines of the Nation;
> – The right to earn enough to provide *adequate food* and clothing and recreation; . . .[3]

FDR had earlier, more abstractly, referred to this right in his 1941 State of the Union speech in which he referred to four freedoms: freedom of speech and expression, freedom to worship, freedom from fear, and freedom from want. He defined freedom from want in this way: "translated into world terms, means economic understandings which will secure to every nation a healthy peacetime life for its inhabitants—everywhere in the world."[4] Franklin's insistence that food is a right has never been pursued in mainstream American political life—neither by Democrats nor by Republicans.

On December 10, 1948, the General Assembly of the United Nations approved the Universal Declaration of Human Rights (UDHR) and it includes the right to food (Article 25): "Everyone has the right to a standard of living adequate for the health and well-being of himself and of his family, including food, clothing, housing and medical care."[5]

Eleanor Roosevelt chaired the committee that drafted the UDHR. It never became a treaty, although it was affirmed by all states and participating non-governmental organizations (NGOs) at the Vienna 1993 World Conference on Human Rights. However, two treaties approved by the General Assembly and that are based on the UDHR were sent out to states for ratification in 1966 and both entered into force in 1976. (All human rights treaties are listed in Chapter 4, Box 4.1.) One of these two is the International Covenant on Civil and Political Rights, and the other is the International Covenant on Economic, Social, and Cultural Rights (ICESCR). Article 11 of the ICESCR highlights that food, along with clothing and housing, is a basic human right: "The States Parties to the present Covenant recognize the right of everyone to an adequate standard of living for himself and his family, including adequate food, clothing and housing, and to the continuous improvement of living conditions."[6]

Article 11 and, especially, the right to food, has been more recently affirmed internationally in 1999,[7] as well as by the UN Special Rapporteur on the Right to Food, who stressed the right to have access to food, and additionally that people have the right to food that is appropriate, according to one's culture:

> the right to have regular, permanent and unrestricted access,
> either directly or by means of financial purchases, to
> quantitatively and qualitatively adequate and sufficient food
> corresponding to the cultural traditions of the people to which the
> consumer belongs, and which ensure a physical and mental,
> individual and collective, fulfilling and dignified life free of fear.[8]

Many state (country) constitutions protect people's right to food or adequate nutrition. For example, Chapter 2 of article 27 of the Constitution of Kenya includes this phrase: "Everyone has the right to have access to . . . sufficient food and water."[9] Regional bodies, including the African Union,[10] the Islamic Organization of Cooperation,[11] and the Organization of American States affirm the right to food in declarations or treaties.[12] To illustrate, in 2012 the General Assembly of the Organization of American States (OAS) approved the "Declaration of Cochabamba on Food Security with Sovereignty in the Americas," which affirms the right to food. Item 7 of the Declaration defines food sovereignty:

> people's right to define their own policies and strategies for the
> sustainable production, distribution, and consumption of food

that guarantee the right to food for the entire population,
respecting their own cultures and the diversity of peasant, fishing,
and indigenous forms of agricultural production, of marketing,
and of management of rural areas, in which women play a
fundamental role.[13]

As noted in Chapter 4, the right to food was incorporated into OAS's 1988 treaty, "Protocol of San Salvador."[14] On July 1, 2013, the African Union unveiled its program to end hunger and malnutrition in all 54 states by 2025.[15] The hope was that achieving this ambitious goal will also protect people and their food supply from calamities accompanying climate change since most African countries are already experiencing unusual heat, droughts, and storms that threaten agriculture.

FOOD IS BASIC AND ELEMENTARY

Without food and water, no person can live long. Studies of hunger strikers have shown that 40 days without food and water is the limit of human endurance.[16] Those of us who live in rich countries take food for granted. Yet food, according to the United Nations Food and Agriculture Organization (UN-FAO), is hardly a given since crops are highly sensitive to extremes of temperature, drought, excessive rain, or, in other words, climate change. Even a 2°C rise in global mean temperature would destabilize current farming systems and would have the potential to endanger food production, "especially the patterns and productivity of crops, livestock, forestry, fisheries, and aquaculture systems."[17]

It is important to note that when scholars and practitioners, human rights activists, and NGO workers say "food is a human right," they mean healthy food *and* clean water. It is imperative to have both. Climate change will adversely affect all this. The World Food Programme estimates that 20% more people will be at risk of hunger by 2050 due to the changing climate, and that clean, potable water will be threatened due to rising sea levels, droughts, and extreme heat.[18] Even so, analyses show that there will be about 9.8 billion people in 2050, which is about 2.2 billion more than there are now, and that will require 70% more food calories than needed now, all the while when it is essential—imperative—to reduce poverty, achieve gender equity, house refugees, and combat extreme weather (droughts, excessive heat, and torrential rains).[19] It won't be easy.

Alas, it may not be the case that everyone will benefit more or less equally from improvements in agricultural productivity. Nor will everyone suffer when there is crop failure. Just as the already privileged enjoy superior economic gains while the already disadvantaged struggle with meager earnings, the privileged will not go hungry while the disadvantaged are at risk of having inferior food or having none at all. This is the case when the comparison is between rich and poor people or between rich and poor countries.[20]

EXAMPLES OF FOOD IMPACTED BY CLIMATE CHANGE

I will explore issues about food and climate change more in depth in later chapters, but this is a good opportunity to provide a background and to begin to highlight some important issues.

Shellfish

The shells of clams, scallops, lobster, oysters, snails, mussels, and shrimp are all affected by ocean acidification just as coral reefs are (although some kinds of algae and sea grasses may benefit from higher CO^2 conditions in the ocean). To be more precise, when CO^2 is absorbed by seawater, a series of chemical reactions occur, resulting in the increased concentration of hydrogen ions. This increase causes the seawater to become more acidic and causes carbonate ions—an important building block of shelled creatures—to be less abundant because carbonate ions are an important building block of structures such as sea shells and coral reefs. It will also become more difficult for these animals to retain a hard shell as acid dissolves their rigid structures.[21]

Fish

Scientists are discovering that food—in the broadest sense of that word—is already impacted by climate change and this is not only irreversible but will get worse. Limiting warming to 1.5°C above pre-industrial levels is, according to most scientists, surely the best option.[22] But in fact, the earth is warming faster than predicted, say, a decade ago. According to the U.S. National Snow and Ice Data Center, the extent of Arctic sea ice has fallen to a new wintertime low.[23] This has profound implications for fish.

As the sea ice melts with warming, so too do ocean waters warm, and this means that fish—all over the world—head toward northern, colder waters. This has been observed by fishermen everywhere, from Bangladesh to Norway.[24] Not only are fish headed North, but so are lobsters, meaning widespread unemployment for fishermen, in Maine, Connecticut, and Rhode Island.[25]

Corn

Corn yields in the central United States have become more sensitive to drought conditions in the past two decades, according to David Lobell, Associate Director of Stanford University's Center on Food Security and the Environment. He predicts that if there were no more changes in temperature sensitivity, crops could lose 15% of their yield within 50 years, or as much as 30% if crops continue the trend of becoming more sensitive over time.[26] The same conclusion has been drawn about U.S. corn by other researchers.[27] Water shortages and warmer temperatures are bad news for corn; according to Lobell, a global rise in temperatures of just 1.8°F would slow the rate of growth by 7%.[28]

Scotch, Coffee, Cherries, Wine Grapes

Many of the things that people enjoy and relish to eat and to drink will be affected. To begin with, Scotch whisky production is already adversely affected by rising temperatures and intense rainstorms.[29] Coffee is also at risk. The Union of Concerned Scientists notes: "Higher temperatures, long droughts punctuated by intense rainfall, more resilient pests and plant diseases—all of which are associated with climate change—have reduced coffee supplies dramatically in recent years."[30] Cherries, of course, are divine-tasty, healthy, nutritious, and lovely to look at as well as to eat. Cherry trees are at risk in Japan and the U.S., two countries that have most prominently supported them. It is also the case that vineyards in which grapes are grown for wine now face serious problems with warming temperatures.[31]

Food Crops in North America, Latin America, Europe, Middle East, Asia, Africa, Pacific Islands

The *Scientific American* has summarized predictions for food crops for different regions. For example, it shows that a spike in average temperatures for North America could hurt its agriculture sector: "as the number of days that are hotter than 30°C (86°F) increases, estimated future harvests of wheat, soybeans and corn could drop by 22 to 49%, depending on the variety of the crop."[32]

Latin America, according to the journal, is particularly affected by El Niño-Southern Oscillation (ENSO), with massive fluctuations in the marine ecosystems off the coasts of Ecuador, Peru, and northern Chile, which are expected to have devastating consequences for fishing, but also for agriculture in the dry corridor of Central America. This reached crisis levels in 2015, with more than 3.5 million people who were food insecure and needed immediate food assistance, healthcare, livelihood recovery, and assistance that would increase their resilience.[33] Farmers in Europe will increasingly face deteriorating conditions—periods of drought as well as periods of intense rain.[34] European governments are encouraging farmers to be self-sufficient and self-reliant, with a shift away from industrial farming to agroecological farming (which will be discussed in a later chapter).

The predictions for the Middle East and Northern Africa are dire. The consensus is that the heat is intolerable now in the hottest months and that some areas will be uninhabitable before 2020. People will flee as climate refugees—that is, of course, unless there are some amazing developments, such as cooled underground homes.[35] The effects of climate change in Asia are already serious and will become more so as over 60% of the population live and/or work on farms.[36] But there is reason to be especially concerned about African agriculture because the continent is heating faster than the rest of the world; intercontinental transportation routes (for food distribution) are not well developed; deserts are more arid; and there have been fewer targeted aid programs from other countries than needed.[37] Most researchers agree that African agriculture faces a calamitous future unless steps are taken soon to reduce dependence on water and to abandon monocropping where it has been introduced. However, by and large pessimism is focused on the large farms and there is greater optimism about small farms on which farmers have relied on indigenous knowledge.[38]

HUNGER

Three United Nations agencies focus on food to reduce worldwide hunger. They have distinct missions while sharing the two goals of increasing worldwide food security and reducing poverty. The Food and Agriculture Organization (FAO) focuses on agriculture, forestry, and fisheries. The World Food Programme (WFP) aims to end hunger and achieve worldwide food security, with the overall goal of achieving global zero hunger. The International Fund for Agricultural Development (IFAD) was founded to support agricultural development projects for food production in developing countries. Box 5.1 summarizes some key facts about hunger.

BOX 5.1 DEFINITIONS AND FACTS ABOUT HUNGER[39]

- Malnourishment or *undernourishment* results from not having enough nutritious foods; *hunger* means not having sufficient food.
- *Stunting* is low height for age, caused by insufficient nutrient intake and frequent infections. It often leads to impaired development.
- *Wasting* is low weight for height, and is the result of acute significant food shortage and/or disease, and can lead to mortality.
- 795 million people do not have enough to eat, although there is enough food to feed everyone on the planet.
- Hunger kills more people than AIDS, malaria and tuberculosis combined.
- About one in eight people, or 13.5% of the overall population, remain chronically undernourished in the developing world, down from 23.4% in 1990–1992.
- As the most populous region in the world, Asia is home to two out of three of the world's undernourished people.
- Sub-Saharan Africa is the region with highest prevalence (percentage of the population) of hunger.
- Hunger or poor nutrition cause nearly half (45%) of deaths in children—3.1 million—each year.
- One in six children in developing countries exhibits wasting.
- In some developing countries, one in three children exhibits stunting.
- 66 million primary school-age children go hungry across the developing world, with 23 million in Africa alone; to reach these 66 million children, WFP calculates that US$3.2 billion is needed.
- There are five types of malnutrition: Protein deficiency (Kwashikor), iron deficiency, vitamin A deficiency, iodine deficiency; and zinc deficiency. Of these, protein deficiency is the most serious, generally leading to death.
- According to United Nations agencies, famine can be declared only when certain levels of mortality, malnutrition and hunger are met. They are: at least 20% of households in an area face extreme food shortages with a limited ability to cope, acute malnutrition rates that exceed 30%, and the death rate exceeds two persons per day per 10,000 persons.

It is important to put the information in Box 5.1 into perspective. There is plenty of food to go around and for everyone on the planet to have enough to eat. Although the world produces 17% more food per person today than 30 years ago, close to a billion people go to sleep hungry every night. The reason for such hunger, according to Jean Ziegler, as the Special Rapporteur on The Right to Food, is converting crops into biofuels, which he contended constitutes "a crime against humanity."[40] Ziegler may have been right, yet not completely. More recently, the United Nations World Food Programme highlights other factors as well. It contends that harmful government policies and market failures can cause mass starvation even when there is enough food because it is too expensive and people simply can't afford it. It also stresses that droughts and natural disasters can exacerbate hunger, and that it is human-driven factors—notably spikes in food prices and conflict—that cause widespread food insecurity, and in extreme cases, famine.[41] In 2017, people in four countries faced famine—Yemen, Nigeria, South Sudan, and Somalia—with war and conflict largely responsible. People elsewhere—South Africa, Zimbabwe, Malawi, and Zambia—are now at risk due to the devastation of crops caused by the invasion of armyworms, brought on by a severe drought—a consequence of climate change.[42] Indeed, we can acknowledge that, for example, the prevailing political and economic crisis in Yemen has led to widespread hunger. Yet confounding the effects of this human-made crisis has been climate change. Yemen has no more water. And this is also the case with Cape Town, which is on the verge of running out of water.[43] In other words, planetary warming and weather-related catastrophes are bound to exacerbate the conditions that lead to hunger, famine and drought.[44]

It is only recently that obesity has become such a major health concern, posing, too, a baffling question about prevalence and epidemiology. Why is it the case that in rich countries the poor have much higher rates of obesity than the rich, while in poor countries the rich have higher rates than the poor?[45] Apparently, the answer is that in rich countries the poor tend to eat what is cheap, such as French fries and fried food, while the rich in poor countries have left the farms and moved to the cities where they are less likely to exercise and more likely to eat Western food.[46]

It is important to understand the tragedy that some people do not have enough to eat when so much food is wasted. Each year, 1.3 billion tons of food—about a third of all that is produced—is thrown out or rots in transit, including about 45% of all fruit and vegetables, 35% of fish and seafood, 30% of cereals, 40–50% of root crops, fruits and vegetables, 20% of oilseeds, meat and dairy products. This means that a significant amount is uneaten and thrown into garbage dumps, and then discarded, it produces methane, which contributes to global warming.[47] It is true that much food is thrown out and this occurs in rich countries, not in poor. Yet in rich countries very little is wasted in transit, while much food is wasted in poor countries while in transit, which, as Jon Mandyck and Eric Schultz explain, is due to inadequate cooling in storage and poor transportation.[48]

THE "GREEN REVOLUTION": A BETRAYAL

Between the 1930s and the late 1960s, agriculture in the developing world was transformed through the "Green Revolution," pioneered by Norman Borlaug, who received the Nobel Peace Prize in 1970 and was credited with saving over a billion people from starvation. It involved the development of high-yielding varieties of cereal grains, expansion of irrigation infrastructure, and modernization of management techniques, with the distribution of hybrid seeds, synthetic fertilizers, and pesticides to farmers. The results were mixed. On one hand, the Green Revolution led to higher yields than was possible with traditional farming techniques, and genetic manipulation made it possible for farmers to grow crops that were disease resistant as well as shorter (that is, not top-heavy).[49]

On the other hand, there were growing objections from farmers from around the world: 1) every genetic intervention was "imperfect" and that led to more intervention, and more imperfection led to even more, which triggered higher costs; 2) fertilizer was expensive; 3) adoption mostly benefited farmers with larger land holdings; 4) it required more water and pesticides, sometimes not easily available or too expensive; 5) it led to a loss of biodiversity; 6) it required fertilizer, which was too expensive for some farmers; 7) it required farmers to buy seeds each year; 8) it depleted water resources; 9) it put traditional varieties of crops at risk;[50] 10) pesticides cause leukemia, Hodgkin's disease, non-Hodgkin's lymphoma, and many other forms of cancer; 11) chemicals introduced in weed killers may cause liver disease, or at least they do so in rats;[51] and 12) chemicals are found to cause excessive weediness in fields.

There have been many criticisms of the Green Revolution, but perhaps the earliest and most important were those of Rachel Carson. She highlighted the dangers of pesticides in her 1962 book *Silent Spring*,[52] which led to the banning of DDT and other pesticides in the United States and the creation of the Environmental Protection Agency. She has had a long-lasting influence, as stressed by Al Gore in his introduction to the 1994 edition of the *Silent Spring*:

> Rachel Carson's influence reaches beyond the boundaries of her
> specific concerns in *Silent Spring*. She brought us back to a
> fundamental idea lost to an amazing degree in modern
> civilization: namely the interconnection of human beings and the
> natural environment.[53]

This interconnection is an ancient idea, which highlights the dignity and skills of farmers, and respects nature.

SHIPPING CHICKENS BACK AND FORTH

Amazingly, things have come full circle. In recent decades, agriculturalists have found or highlighted innovative ways of restoring the interconnection between human beings and the natural environment. In particular, in the next chapter I will highlight the Food Sovereignty Movement, Agroecology, and the Local Food Movement. But first, I would like to mention here one of the most bizarre practices of food production that exemplifies industrial agriculture.

It is likely that the chicken leg you ate last time came from a chicken raised in the United States but was butchered and cut up at a processing plant in China and shipped back to the United States for you to eat. This is the arrangement struck between the United States and China, and finalized on August 30, 2013.[54] For one thing, this arrangement flies in the face of long-established—really ancient—farming practices in which the farmer and her or his animals have remarkably close—yes—relations, which is to say that although farmers may have slaughtered the animals, they took great care of them while they were alive. Treating animals, even birds, like mere products or objects is highly objectionable to many farmers.

For another thing, whole chickens and chicken parts transported a total of around 14,000 miles, whether by air or sea, contributes to planetary warming. Aircraft emit carbon dioxide (CO_2), methane, nitrous oxide, hydrofluorocarbons (HFCs), perfluorocarbons (PFCs), and sulfur hexafluoride (SF_6), all of which contribute to GHG,[55] and, likewise, cargo ships also contribute to GHG—1.12 billion tons of CO_2 per year, which is roughly 4% of the world's overall output of CO_2.[56] Fish are also shipped, but the majority are chickens. This export–reimport agreement between the Food Safety and Inspection Service of the US Department of Agriculture and the Peoples Republic of China was finalized on August 29, 2013 after a comprehensive four-year audit of the relevant sections of the food safety system in China, and was finalized on March 4 to March 19, 2013.[57]

The agreement is not clear as to how the chickens and fish are shipped, but let's imagine it is by air. That would mean that each chicken and its parts and each fish are air transported about 15,000 miles. It is important to keep in mind that airplanes account for about 5% of global CO_2.[58] It wouldn't take long by air, but it would take a huge toll on the environment. Shipping by sea would take about 30 days each way. Say the ship went from New York (with whole chickens) to Guangzhou and from Guangzhou to New York (with chicken parts) for a total of about 60 days. According to a United Nations' International Maritime Organization (IMO) report, international shipping by boat in 2007 was responsible for 3.3% of global CO_2, which is predicted to rise to 6% of the total by 2020.[59]

Yes, this practice, whether by air or ship, is highly objectionable because transportation over such long distances contributes to global warming, and, besides, the purpose of doing this is highly questionable. It also casts aspersion on American agricultural practices and the way that food—and nature—has been commodified, or turned into purely marketable items, bought and sold, available mostly to those with money. Also, it is the concept of what is live—from carrots to cows—is merely valuable as a commercial product—that is, vegetables are not grown, they are produced, and chickens are not raised, they are mass produced. Vegetables and chickens (and pigs) are produced like widgets or gadgets, to be bought and sold, which is to say that food becomes produced for a profit, not nourishment. How much? How much do I pay? If I can't pay, I can't eat.

Chickens (and animals) illustrate what is a much broader practice of commoditization. Take dairy cows. Typically, dairy cows are reared and treated as if they are machines.[60] When a female cow gives birth, she will bellow and scream, sometimes for days, because her baby calf is taken from her. This is profitable because the milk intended for her baby is taken and sold, and she will continue to produce milk. If the calf is male, he is likely to be killed and sold for veal. If the calf is female, her fate is sealed as a milk-producing and baby-producing machine. *Rollingstone* has this account about pigs:

> Smithfield's pigs live by the hundreds or thousands in warehouse-like barns, in rows of wall-to-wall pens. Sows are artificially inseminated and fed and delivered of their piglets in cages so small they cannot turn around. Forty fully grown 250-pound male hogs often occupy a pen the size of a tiny apartment. They trample each other to death. There is no sunlight, straw, fresh air or earth.[61]

As Americans, we have become used to the commodification and exploitation of animals.[62] Of course, it looks innocuous since the publicized and stated purpose is to preserve nature, but the result is to preserve nature in order to price and monetize it. Increasingly, investors in land are finding that selling rather than preserving land brings them big returns.

KEY PROBLEMS

Big business dominates our global food system. A small handful of large corporations, including Nestlé, General Mills, and Kellogg, control much of the worldwide production, processing, distribution, marketing and retailing of food. This concentration of power enables big businesses to wipe out competition and dictate tough terms to their suppliers. It forces farmers and consumers into poverty. Under this system, around a billion people go hungry and around two billion are obese or overweight.

Let's start with the tenets of neoliberalism in order to show how they fail to square with the well-being of people and society's advance under the conditions that will accompany climate warming. Neoliberalism privileges personal gain and profit, whereas the views reflected in the Bolivian and Ecuadorean constitutions highlight philosophies of accommodation and cooperation, with the understanding that Mother Earth is all-encompassing and inherently gentle. But is now at risk, with ferocious storms, floods, hot weather, melting poles, warming oceans, and exceedingly hot temperatures. She deserves protection, not abuse. Surely, according to this view, privatization of Mother Earth will not work, since privatization will lead to the exploitation of nature, precisely the wrong approach in time of duress. Privatization entails rolling back the rules, liberating the "free market" from government control, reducing wages by deunionizing workers, and eliminating workers' rights that had been won over after many years of struggle. It also involves the elimination of price controls, and, instead, giving total freedom of movement for capital, goods and services. To convince us that this is good for us, politicians say that "an unregulated market is the best way to increase economic growth and benefit everyone." In fact, the idea is to grow the economy for private gain by cutting public expenditures for social service, education, healthcare and reducing public expenditures for roads, bridges, and schools. This accompanies deregulation to save money, and to abolish environmental and safety rules and regulations.

It is quite clear that apart from the sheer unfairness of a system that privileges private wealth accumulation over people's well-being and the environment, a radically deregulated economy is the worst thing to do at a time of global warming. To stem global warming, we need to cooperate and to get away from agricultural practices that are based on pesticides and monoculture farming, and to promote farming practices that preserve the soil and water. The defining trait of monoculture farming is that the aim is to maximize profits and yields, while producing healthy food for everyone is not an objective. Increasingly, monoculture is being criticized for many reasons: it is susceptible to disease, it needs excessive amounts of water, it degrades the soil, and uses chemicals and fertilizers, and thereby contributes to climate change.

Another of the many worrying aspects of monoculture/industrial farming is a dramatic decline in the genetic variety or diversity of crops. According to the international Food and Agriculture Organization:

> More than 90 percent of crop varieties have disappeared from farmers' fields; half of the breeds of many domestic animals have been lost. In fisheries, all the world's 17 main fishing grounds are now being fished at or above their sustainable limits, with many fish populations effectively becoming extinct. Loss of forest cover, coastal wetlands, other "wild" uncultivated areas, and the destruction of the aquatic environment exacerbate the genetic erosion of agrobiodiversity.[63]

CONCLUSIONS: CLIMATE CHANGE, AGRICULTURE, AND TRUMP

Now we come to climate change, the role that industrial farming has played in accelerating climate change and Trump's agricultural policies, as promoted by Sonny Purdue, Secretary of Agriculture. Industrial agriculture is responsible for over 50% of greenhouse gas emissions through intensive use of agrochemicals, toxins, fossil energy, freight land grabbing, and forest degradation through plantations, mining, logging etc. Multinational agricultural corporations include Monsanto, Dow, BASF, Bayer, Syngenta and DuPont. They control the global seed, pesticide and agricultural biotechnology markets. They have displaced farmers and greatly reduced biodiversity. Sonny Purdue is an advocate of industrial agriculture in the U.S. Trump's motto "America First" finds expression in, among other arenas, a food policy that promotes trade practices that are not sustainable and an agricultural policy that will increase greenhouse gases. The rest of the world is appalled.

NOTES

1   Belarus constitution, section 2, article 21. Constituteproject: Fiji constitution, Article 36: Constitute project. Available at: www.constituteproject.org/search?lang =en&key=standliv; www.constituteproject.org/constitution/Fiji_2013?lang=en

2   See: Asbjorn Eide, Wenche Barth Eide, Susantha Goonatilake, Joan Gussow, and Omawale (Eds.) *Food as a Human Right.* Tokyo: United Nations University, 1984; Joan Goldstein, *Clean Food and Water: The Fight for a Basic Human Right.* New York: Plenum Press, 1990; George Kent (Ed.), *Global Obligations for the Right to Food.* Lanham, MD: Rowman & Littlefield, 2008.

3   Franklin D. Roosevelt, "State of the Union Message to Congress." Available at: www.fdrlibrary.marist.edu/archives/address_text.html (emphasis added).

4   National Endowment for the Humanities, "FDR's Four Freedoms Speech." Available at: https://edsitement.neh.gov/lesson-plan/fdrs-four-freedoms-speech-freedom-fireside

5   United Nations. Universal Declaration of Human Rights. Available at: www. un.org/en/universal-declaration-human-rights/

6   International Covenant on Economic, Social and Cultural Rights. Adopted and opened for signature, ratification and accession by General Assembly resolution 2200A (XXI), December 16, 1966; entry into force January 3, 1976, in accordance with article 27. Available at: www.ohchr.org/EN/ProfessionalInterest/Pages/CESCR.aspx

7   The right to adequate food has been subsequently affirmed as indispensable for the inherent dignity of the person and for the fulfillment of other human rights. See "General Comment No. 12. The Right to Adequate Food (Article 11 of the Covenant)," UN Committee on Economic, Social and Cultural Rights, May 12, 1999. Available at: www.refworld.org/docid/4538838c11.htm; also see FIAN International. The Right to Adequate Food. Available at: www.fian.org/what-we-do/issues/right-to-food/

8   Special Rapporteur on the Right to Food. Available at: www.ohchr.org/EN/Issues/Food/Pages/FoodIndex.aspx

9   Available at: www.constituteproject.org/search?lang=en&key=standliv

10   African Union. Protocol to the African Charter on Human and People's Rights on the Rights of Women in Africa. Avilable at: www.achpr.org/instruments/women-protocol/

11   Organization of Islamic Cooperation. The Statute of the Organization for Food Security. Available at: www.oic-oci.org/upload/pages/conventions/en/STATUTE-IOFS-ENG-FINAL.pdf

12   Organization of American States. Coherent Food Security Responses. Available at: www.fao.org/righttofood/our-work/current-projects/rtf-global-regional-level/oas/en/

13   General Assembly. Forty-Second Regular Session. Declaration of Cochabamba on "Food Security with Sovereignty in the Americas." Available at: www.fao.org/fileadmin/templates/righttofood/documents/project_m/oas/declaration-cocha bamba-Food-Security-Sovereignty-Americas_en.pdf

14   Additional Protocol to the American Convention on Human Rights in the Area of Economic, Social and Cultural Rights "Protocol of San Salvador." Available at: www.oas.org/juridico/english/treaties/a-52.html

15 African Union. High level meeting. Available at: http://pages.au.int/category/special-pages/endhunger

16 *Scientific American*, "How long can a person survive without food?" Available at: www.scientificamerican.com/article/how-long-can-a-person-sur/

17 UN Food and Agriculture Organization (FAO). FAO's Work on Climate Change: United Nations Conference on Climate Change 2016, p. 10. Available at: www.fao.org/3/a-i6273e.pdf

18 World Food Programme. *Climate Change*. Available at: www.wfp.org/node/12099. Also see UN Water. *Water and Climate Change*. Available at: www.unwater.org/topics/water-and-climate-change/en/ World Resources Institute. *Water*. Available at: www.wri.org/our-work/topics/water

19 World Resources Report 2013–2016: *Creating a Sustainable Food Future*. Available at: www.wri.org/our-work/project/world-resources-report/world-resources-report-2013-2016-creating-sustainable-food

20 Genevieve Lavoie-Mattieu, "Positive outlook for agricultural prices not for world's poorest." *Inter Press Service*, July 25, 2014. Available at: www.ipsnews.net/2014/07/positive-outlook-for-agricultural-prices-but-not-for-worlds-poorest/

21 Ocean Acidification International Coordination Centre. "Ocean acidification crumbling the shells of the sea." Available at: https://news-oceanacidification-icc.org/2014/11/20/ocean-acidification-crumbling-the-shells-of-the-sea/

22 Delphine Deryng, "What's on the science agenda for the IPSS Special Report on 1.5C in agriculture." *Climate* Analytics. January 16, 2017. Available at: http://climateanalytics.org/blog/2017/whats-on-the-science-agenda-for-the-ipcc-special-report-on-1-5c-in-agriculture.html

23 National Snow and Ice Data Center. "Arctic sea ice maximum at record low for third straight year." Available at: https://nsidc.org/arcticseaicenews/2017/03/arctic-sea-ice-maximum-at-record-low/

24 Robin McKie, "How warming seas are forcing fish to seek new waters." *The Guardian*, January 8, 2017. Available at: www.theguardian.com/environment/2017/jan/08/fish-ocean-warming-migration-sea

25 Patrick Whittle, "Lobsters head north to cold." *Guelph Mercury Tribune*, August 18, 2015. Available at: www.guelphmercury.com/news-story/5803693-lobsters-head-north-to-cold/

26 David B. Lobell, Michael J. Roberts, Wolfram Schlenker, Noah Braun, Bertis B. Little, Roderick M. Rejesus, and Graeme L. Hamme, "Greater sensitivity to drought accompanies maize yield increase in the U.S. Midwest." *Science*, 344(6183): 516–519, May 2, 2014. DOI: 10.1126/science.1251423. Available at: http://science.sciencemag.org/content/344/6183/516. Twilight Greenaway, "Eight foods you're about to lose due to climate change." *The Guardian*, October 29, 2014. Available at: www.theguardian.com/vital-signs/2014/oct/29/diet-climate-maple-syrup-coffee-global-warming

27 Becky Beyers, Department of Soil, Water and Climate, University of Minnesota, "Changing climate will reduce corn production." Available at: www.swac.umn.edu/changing-climate-corn-production

28 Greenaway. "Eight foods you're about to lose due to climate change." *The Guardian*, October 29. 2014. Available at: www.theguardian.com/vital-signs/2014/oct/29/diet-climate-maple-syrup-coffee-global-warming

29   ScotchWhisky.com. "How will climate change affect scotch?". Available at: https://
     scotchwhisky.com/magazine/features/13034/how-will-climate-change-affect-
     scotch/. Greenaway, "Eight foods you're about to lose due to climate change."
     The Guardian, October 29, 2014. Available at: www.theguardian.com/vital-signs/
     2014/oct/29/diet-climate-maple-syrup-coffee-global-warming

30   Union of Concerned Scientists. "Coffee and climate. What's brewing with climate
     change?" Available at: www.ucsusa.org/global_warming/science_and_impacts/
     impacts/impacts-of-climate-on-coffee.html#.WNZfkGe1uUk. Climate Institute.
     "A brewing storm." August 2016. Available at: www.climateinstitute.org.au/verve/_
     resources/TCI_A_Brewing_Storm_FINAL_WEB270916.pdf

31   Cornelis van Leeuwen and Phillippe Darriet, "The impact of climate change on
     viticulture and wine quality." *Journal of Wine Economics*, 11(1), 2016. Available at:
     www.wine-economics.org/aawe/wp-content/uploads/2016/06/Vol 11-Issue01-
     The-Impact-of-Climate-Change-on-Viticulture-and-Wine-Quality.pdf

32   Kavya Balaraman, "U.S. crop harvest could suffer with climate change." *Scientific
     American*, January 20, 2017. Available at: www.scientificamerican.com/article/
     u-s-crop-harvests-could-suffer-with-climate-change/. L. Ziska, A. Crimmins,
     A. Auclair, S. DeGrasse, J.F. Garofalo, A.S. Khan, I. Loladze, A.A. Pérez de León,
     A. Showler, J. Thurston, and I. Walls, *2016: Food Safety, Nutrition and Distribution*.
     U.S. Global Change Research Program, Washington, DC. Available at: https://
     health2016.globalchange.gov/food-safety-nutrition-and-distribution. Environmental
     Protection Agency, "Climate Change Impacts." Available at: www.epa.gov/climate-
     impacts/climate-impacts-agriculture-and-food-supply#crops

33   Intergovernmental Panel on Climate Change. "The regional impacts of climate
     change." Available at: www.ipcc.ch/ipccreports/sres/regional/index.php?idp=123.
     United Nations Office for the Coordination of Humanitarian Affairs, "El Niño
     in Latin America and Caribbean." Available at: www.unocha.org/el-nino-latin-
     america-caribbean

34   European Commission. "Agricultural and climate change." Available at: https://ec.
     europa.eu/agriculture/climate-change_enhttps://ec.europa.eu/agriculture/climate-
     change_en

35   VOA News. "Climate change could make parts of Middle East uninhabitable."
     Available at: www.voanews.com/a/mht-climate-change-could-make-parts-of-
     middle-east-uninhabitable/3311862.html

36   Asian Development Bank. "Agriculture and climate change." Available at: www.
     adb.org/news/photo-essays/agriculture-and-climate-change

37   For a summary, see Lahouri Bounoua, "Climate change is hitting African farmers
     the hardest of all." *The Conversation*. May 12, 2015. Available at: http://theconversa
     tion.com/climate-change-is-hitting-african-farmers-the-hardest-of-all-40845

38   A. Nyong, F. Adesina, and B. Osman Elash, "The value of indigenous knowledge
     in climate change mitigation and adaptation strategies in the African Sahel."
     Mitigation and Adaptation Strategies for Global Change (2007) 12: 787–797. DOI
     10.1007/s11027-007-9099-0. Available at: www.researchgate.net/profile/Anthony_
     Nyong/publication/46537001_The_Value_of_Indigenous_Knowledge_in_Climate
     _Change_Mitigation_and_Adaptation_Strategies_in_the_African_Sahel/links/5547
     6f460cf26a7bf4d906e6/The-Value-of-Indigenous-Knowledge-in-Climate-Change-
     Mitigation-and-Adaptation-Strategies-in-the-African-Sahel.pdf. Climate change in
     Southern Africa. Available at: http://media.csag.uct.ac.za/faq/qa_3impacts.html

39 Sources: World Hunger.Org. "World hunger and poverty facts and statistics." Available at: www.worldhunger.org/2015-world-hunger-and-poverty-facts-and-statistics/. United Nations Food and Agricultural Organization. *The State of Food Insecurity in the World 2015.* Available at: www.fao.org/hunger/en/. World Food Program. *Hunger Statistics.* Available at: www.kidsagainsthungersiouxfalls. org/images/stories/kah_pdf/resources/hunger_statistics_2015.pdf; https://wfp.org/ hunger/stats. Medscape, September 24, 2016. "Protein energy malnutrition." Available at: http://emedicine.medscape.com/article/1104623-overview#a6. World Food Program. *Types of Malnutrition.* Available at: www.wfp.org/hunger/malnutri tion. Food and Agriculture Organization. "Hunger Facts." Available at: www.fao. org/about/meetings/icn2/toolkit/hunger-facts/en. UN News Centre. "When a food crisis becomes a famine." Available at: www.un.org/apps/news/story.asp?News ID=39113#.WLwPfvLmyZY

40 United Nations News Center. "UN independent rights expert calls for five-year freeze on biofuel production." October 26, 2007. Available at: www.un.org/apps/ news/story.asp?NewsID=24434#.WLwZk_LmyZY

41 Available at: www.wfpusa.org/articles/crash-course-drought-and-famine/

42 Associated Press in Harare, *The Guardian.* "Plague of armyworms threatens to strip southern Africa of key food crops." Available at: www.theguardian.com/world/ 2017/feb/14/southern-africas-food-crops-under-threat-from-fall-armyworm-invasion

43 *National Geographic.* "Cape Town is running out of water." Available at: https:// news.nationalgeographic.com/2018/02/cape-town-running-out-of-water-drought-taps-shutoff-other-cities/

44 Jean Drèze and Amartya Sen pointedly note: "It is not so much that there is no law against dying of hunger. That is, of course, true and obvious. It is more that the legally guaranteed rights of ownership, exchange and transaction delineate economic systems that can go hand in hand with some people failing to acquire enough food for survival." Jean Drèze and Amartya Sen, *Hunger and Public Action.* Oxford, UK: Clarendon Press, 1989, p. 20.

45 James A. Levine, "Poverty and obesity in the U.S." *Diabetes.* November, 2011, 60(11): 2667–2668. Published online October 17, 2011. Doi: 10.2337/db11-1118. Barry M. Popkin, "The nutrition transition in the developing world," pp. 43–56 in Simon Maxwell and Rachel Slater (Eds.) *Food Policy Old and New.* Oxford, UK: Blackwell, 2004.

46 D.J. Hoffman. "Obesity in developing countries." UN FAO. Available at: www. fao.org/docrep/003/y0600m/y0600m05.htm/

47 UN Food and Agriculture Organization. Global Initiative on Food Loss and Waste Reduction, 2015. Available at: www.fao.org/3/a-i4068e.pdf

48 John M. Mandyck and Eric B. Schultz, "Food foolish: The hidden connection between food waste, hunger and climate change." Farmington, CT: Carrier Corporation, 2015.

49 Howard D. Leathers and Phillips Foster. *The World Food Problem: Toward Ending Undernutrition in the Third World.* Boulder, CO: Lynne Rienner, 2009, pp. 213–236.

50 Ibid. Raj Patel, *Stuffed and Starved: The Hidden Battle for the World Food System.* Brooklyn, NY: Melville House, 2012, pp. 129–136. John M. Mandyck and Eric B Schultz, "Food foolish: The hidden connection between food waste, hunger and climate change." Farmington, CT: Carrier Corporation, 2015. Food and Agriculture

Organization. World Food Summit 1996. Towards a New Green Revolution. Available at: www.fao.org/docrep/x0262e/x0262e06.htm#TopOfPage

51   Robin Mesnage, George Renney, Gilles-Eric Séralini, Malcolm Ward and Michael N. Antoniou, "Multiomics reveal non-alcoholic fatty liver disease in rats following chronic exposure to an ultra-low dose of Roundup herbicide." *Scientific Reports*, 7, January 9, 2017. Doi: 10.1038/srep39328

52   Rachel Carson, *Silent Spring*. Cambridge, MA: Houghton Mifflin, 1962. For an analysis of Carson's contributions, see Joan Goldstein, *Demanding Clean Food and Water: The Fight for a Basic Human Right*. New York: Plenum Press, 1990.

53   Al Gore, "Introduction." Rachel Carson, *Silent Spring*. Boston, MA: Houghton Mifflin, 1994, p. xxvi.

54   United States Department of Agriculture. Letter from Director, International Equivalence Office, United States, to Deputy Director, General Inspection and Quarantine, Beijing, August 30, 2013, and Final Report of the Audit of the Food Safety System Governing the Production of Processed Poultry Intended for Export to the United States of America. Available at: www.fsis.usda.gov/wps/wcm/connect/c3dab827-151d-4373-917f-139db6a2466d/China_2013_Poultry_Processing.pdf?MOD=AJPERES

55   Intergovernmental Panel on Climate Change. Aviation and the Global Atmosphere. 6.1. How do aircraft cause climate change? Available at: www.ipcc.ch/ipccreports/sres/aviation/index.php?idp=65

56   Michael Hopkin, "Ships' greenhouse emissions revealed." *Nature*, February 13, 2008. Available at: www.nature.com/news/2008/080213/full/news.2008.574.html

57   Final Report of an Audit Conducted in the People's Republic of China. March 4 through 19, 2013. FSIS Audit of the Food Safety System Governing the Production of Processed Poultry Intended for Export to the United States of America. Available at: www.fsis.usda.gov/wps/wcm/connect/ed782de3-82e1-4298-aac9-14da84d1ebd2/2013_China_Poultry_Slaughter_FAR.pdf?MOD=AJPERES; USDA. FSIS. FSIS Reaffirms Equivalence of China Poultry Processing System, August 20, 2013. Available at: www.fsis.usda.gov/wps/portal/fsis/newsroom/news-releases-statements-transcripts/news-release-archives-by-year/archive/2013/nr-08302013-01; also see: www.fsis.usda.gov/wps/portal/fsis/newsroom/news-releases-statements-transcripts/news-release-archives-by-year/archive/2016/faq-china-030416

58   Elisabeth Rosenthal, "Your biggest carbon sin may be air travel." *New York Times*, January 26, 2013. Available at: www.nytimes.com/2013/01/27/sunday-review/the-biggest-carbon-sin-air-travel.html

59   Seas at Risk. "Shipping and climate change." Available at: www.seas-at-risk.org/issues/shipping/shipping-and-climate-change.html

60   *Viva!* "The dark side of dairy." Available at: www.whitelies.org.uk/sites/default/files/dark_side_of_dairy_report_2014.pdf. Chas Newkey-Burden, "Dairy is scary." March 30, 2017. Available at: www.theguardian.com/commentisfree/2017/mar/30/dairy-scary-public-farming-calves-pens-alternatives

61   *Rollingstone.* "Boss hog: The dark side of America's top pork producer." Available at: www.rollingstone.com/culture/news/boss-hog-the-dark-side-of-americas-top-pork-producer-20061214

62   Although seemingly altruistic commodification, mitigation banking commodifies nature by investing and speculating in natural resources. See: Vikram Jhwar, Understanding the basics of mitigation banking, March 16, 2015. Available at:

www.aljazeera.com/programmes/specialseries/2015/11/pricing-planet-1511191
11851963.html. Aljazeera, "Pricing the planet." Available at: www.aljazeera.com/
programmes/specialseries/2015/11/pricing-planet-151119111851963.html
63   UN Food and Agriculture Organization. "What is happening to agrobiodiversity?"
Available at: www.fao.org/docrep/007/y5609e/y5609e02.htm

CHAPTER 6

The Peasant Farmers were Right All Along

The most comprehensive critique of the Green Revolution—industrial agriculture—is that it helped to spawn climate change and continues to exacerbate and magnify its effects. First, industrial agriculture relies on fossil fuels; second, the crops it supports and produces are not resilient to changes in weather and heat which accompany climate change; and, third, it has detrimental effects such as soil depletion and has led to the decline of biodiversity. Moreover, industrial agriculture or agribusiness has been taken over by capitalists to maximize profits, not quality food.

BIOLOGICAL DIVERSITY

At the 1992 Earth Summit in Rio de Janeiro, world leaders agreed on a comprehensive strategy for "sustainable development," meeting people's needs while ensuring that we leave a healthy and viable world for future generations. One of the key agreements adopted at Rio was the Convention on Biological Diversity, which sets out commitments for ensuring and maintaining the world's ecological resources as people, businesses, and countries go about the business of economic development. The Convention establishes three main goals: the conservation of biological diversity, sustainability, and the fair and equitable sharing of the benefits from the use of genetic resources. No, it does not explicitly condemn agribusiness and the methods of the Green Revolution, but it clearly supports the principles advocated by peasants advancing the food sovereignty movement (which I will explain below).[1] Box 6.1 provides extracts from the Convention.

BOX 6.1 CONVENTION ON BIOLOGICAL DIVERSITY (1992): 196 PARTIES (EXTRACTS)

Conscious of the intrinsic value of biological diversity and of the ecological, genetic, social, economic, scientific, educational, cultural, recreational and aesthetic values of biological diversity and its components,

Conscious also of the importance of biological diversity for evolution and for maintaining life sustaining systems of the biosphere,

Affirming that the conservation of biological diversity is a common concern of humankind,

Reaffirming that States have sovereign rights over their own biological resources,

Noting that it is vital to anticipate, prevent and attack the causes of significant reduction or loss of biological diversity at source,

Noting also that where there is a threat of significant reduction or loss of biological diversity, lack of full scientific certainty should not be used as a reason for postponing measures to avoid or minimize such a threat,

Noting further that the fundamental requirement for the conservation of biological diversity is the in-situ conservation of ecosystems and natural habitats and the maintenance and recovery of viable populations of species in their natural surroundings,

Recognizing the close and traditional dependence of many indigenous and local communities embodying traditional lifestyles on biological resources, and the desirability of sharing equitably benefits arising from the use of traditional knowledge, innovations and practices relevant to the conservation of biological diversity and the sustainable use of its components,

Recognizing also the vital role that women play in the conservation and sustainable use of biological diversity and affirming the need for the full participation of women at all levels of policy-making and implementation for biological diversity conservation,

Noting in this regard the special conditions of the least developed countries and small island states,

Acknowledging that substantial investments are required to conserve biological diversity and that there is the expectation of a broad range of environmental, economic and social benefits from those investments,

Recognizing that economic and social development and poverty eradication are the first and overriding priorities of developing countries.

Source: United Nations. Convention on Biological Diversity. Available at: www.cbd.int/convention/articles/default.shtml?a=cbd-00

AGROECOLOGY

At the 1996 World Food Summit, participants referred to the negative aspects of the Green Revolution, such as loss of genetic diversity, deforestation, excess use of fertilizers and pesticides, soil erosion, and land degradation. The new paradigm that was emerging—although not named as yet—focused on biological diversity, the retention of soil nutrients, preservation of the environment, and the role of small farmers, especially women farmers.

Around 2010, the term "agroecology" began to be used to describe this new paradigm, probably mentioned first most prominently in a report to the UN General Assembly submitted on behalf of the UN Human Rights Council by Olivier de Schutte, the Special Rapporteur on the Right to Food,[2] who defined agroecology in the following way:

> Agroecology is based on applying ecological concepts and principles to optimize interactions between plants, animals, humans and the environment while taking into consideration the social aspects that need to be addressed for a sustainable and fair food system. By building synergies, agroecology can support food production and food security and nutrition while restoring the ecosystem services and biodiversity that are essential for sustainable agriculture. Agroecology can play an important role in building resilience and adapting to climate change.
>
> Agroecology looks for local solutions and linkages with the local economy and local markets, and keeps farmers in the field with improved livelihoods and a better quality of life.[3]

In other words, in advocating agroecology, the Special Rapporteur contended that agroecology is not consistent with industrial agriculture and instead highlights local practices, sustainability, farmers' well-being, biodiversity, and nutrition. The two major UN food agencies also contended that agroecology practices are better adapted to climate change than industrial agriculture.[4]

A CONVERGENCE: SLOW FOOD, AGROECOLOGY, AND BIODIVERSITY

The Slow Food Movement is to consumers what the Agroecology Movement is to growers. Agroecology practices privilege fulsome and sustainable methods of farming, without monocropping, GMOs and pesticides, with an emphasis on biodiversity. The Slow Food Movement, which was founded in Italy by Carlo Petrini in 1986, is opposed to fast food and GMOs, defends the right to food, and promotes biodiversity as well as the diversity of cultures. This statement helps to capture the philosophy of the Slow Food Movement:

> Meadows, woods, gardens, hamlets, villages, towns and cities that respect Nature—they form the daily landscape that we like the best, the one we would like to see every day and love to visit when we are on our travels. The landscape and its beauty are treasures that help us to live better, that make us feel well. They make existence more pleasant and they enhance the pride we feel for our land. Through food we thus have the opportunity to ensure that beauty surrounds us all the time and that future generations can enjoy it too. Beauty is not an option nor a luxury, nor is it antithetical to human progress. In industrial societies too, much beauty has been squandered for the sake of a misguided idea of "progress." Victims of "progress" are also the many agrarian societies that suffer neglect and dereliction or an excessive intensification of farming activities. Beauty has disappeared from their countryside.
>
> Forecasts all seem to agree on the fact that, in 2050, there will be 9 billion people sharing the planet. Considering that today (with a world population of 7 billion) a billion people do not eat adequately, the prospects look gloomy. The most disparate and "authoritative" voices are increasingly stressing the fact that, to feed everyone, it will be necessary to increase productivity by 70% (with cultivated land decreasing in the meantime).
>
> Hence the rush to genetically manipulate seeds to create hyperproductive plant species. Hence the idea of feeding meat animals on antibiotics and hormones to make them grow in half the normal time. Hence the inevitable destruction of forests to obtain arable land (which nonetheless loses its fertility in the space of a few seasons anyway). In short, who can worry about biodiversity, animal well-being and climate change when people are dying or risk dying of hunger?
>
> But there is an element missing in this analysis—often left unsaid more in bad faith than out of shallowness—that cannot help but leave us with our stomachs knotted: namely, that today the earth produces food for 12 billion people. Forty percent of all food produced is wasted and turns to waste without even getting near to the table.[5]

This statement celebrates a holistic understanding of community, nature, and food, while along the lines of the advocates for seed sovereignty and agroecology it opposes GMOs and the manipulation of genes. It is important to point out that not only have the FAO and WFP come out opposing agribusiness and industrial agriculture, but so, perhaps surprisingly, did the US Department of Agriculture.[6] Likewise, the Slow Food Movement is similar to agroecology. Both are holistic and integrative, affirming community values and traditional farming practices. Peasant farmers also advocate Slow Food or agroecology, and they call it "Food Sovereignty."

FOOD SOVEREIGNTY

La Via Campesina has been the voice for indigenous farmers and represents more than 200 million farmers, describing itself as "the international movement which brings together millions of peasants, small and medium-size farmers, landless people, women farmers, indigenous peoples, migrants and agricultural workers from around the world." It defends small-scale sustainable agriculture as a way to promote social justice and dignity, as well as quality food. La Via Campesina strongly opposes corporate-driven agriculture and transnational companies that are destroying people and nature.[7]

La Via Campesina coined the term "Food Sovereignty." It is the right of peoples to healthy and culturally appropriate food produced through sustainable methods and their right to define their own food and agriculture systems while being opposed to genetically modified seeds, the corporatization of agriculture, monocultural practices, bioengineering, and threats to biodiversity.[8] In other words, advocates of Food Sovereignty favor holistic farming that does not rely on pesticides and herbicides, and genetically modified seeds. Interestingly, this is the position of the UN's Food and Agriculture Organization that defends food sovereignty because the corporatization of agriculture, monocropping and genetic modification of seeds threaten biodiversity, and therefore the quality and quantity of the food supply.[9] At the 1996 World Food Summit, coordinated by the Food and Agricultural Organization, a conclusion was:

> But steps must also be taken to avoid the social and environmental damage caused by a wholesale shift to monoculture production. A new Green Revolution will need to combine modern technology, traditional knowledge and an emphasis on farming, social and agro-ecological systems as well as yields.[10]

There is, at this very moment, a global agricultural revolution. Its leaders are peasant farmers. In fact, they started doing it long ago, but now the rest of us are beginning to join them. La Via Campesina calls it "agroecology." It is remarkable that so few Americans know about it, although a few U.S. university horticultural departments are beginning to teach it. Now, notably led and organized by peasant farmers with La Via Campesina at the helm, it simply highlights what farmers have long been doing. They call it agroecology. Advocates oppose large-scale, industrial farming, which employs monocropping, chemical-based industrial or commercial practices, and the extensive use of herbicides and pesticides. Practitioners of agroecology reject corporate mega-farms and overcrowded conditions prevalent in large cattle and hog farms. First, I recall an interesting and important story that I mentioned in Chapter 5.

Let us contrast two quite different visions—indeed, two fundamentally different ones—and then I will bring in a third. One is mitigation banking, where nature is bought and sold presumably to protect it, which easily veers toward the appropriation and exploitation of nature. "Sure, I'm a little short of cash this month. I'll just sell a few trees that I have banked." The point I want to bring home is that trees are not currency or investments. They should not be the object of speculation. Yes, of course, we buy a tree to plant it, but not as an investment. Relevant is the way that two countries—Bolivia and Ecuador—portray the natural world. The Bolivian and Ecuadorean constitutions describe nature in reverential terms, as "Mother Earth."[11] Here is the section of the Bolivian Constitution that is relevant:

> In ancient times mountains arose, rivers moved, and lakes were formed. Our Amazonia, our swamps, our highlands, and our plains and valleys were covered with greenery and flowers. We populated this sacred Mother Earth with different faces, and since that time we have understood the plurality that exists in all things and in our diversity as human beings and cultures. Thus, our peoples were formed, and we never knew racism until we were subjected to it during the terrible times of colonialism.

The following is Ecuador's statement, which is shorter: "*Celebrating nature, the Pacha Mama (Mother Earth), of which we are a part and which is vital to our existence.*"

Both constitutions led to laws that prohibit buying, selling or trading nature. She is us and we are her. She ensures our beautiful world, our diversity, our racial groups and identities, our world of which we are a part. This world view stands in opposition to neoliberalism, the dominant economic paradigm in the U.S. Just to note again, in American factory farms, animals are packed so tight they can barely breathe.[12]

THE ORIGINS OF FOOD SOVEREIGNTY AND AGROECOLOGY

Farmers from all over the world rejected monocultural/industrial farming, but it was Viva la Campesina in 1966 that first publically highlighted the dangers of monoculture agriculture and to advance food sovereignty, which it defines as follows:

> Food sovereignty is the right of peoples to healthy and culturally appropriate food produced through ecologically sound and sustainable methods, and their right to define their own food and agriculture systems. It puts those who produce, distribute and consume food at the heart of food systems and policies rather than the demands of markets and corporations. It defends the interests and inclusion of the next generation. It offers a strategy to resist and dismantle the current corporate trade and food regime, and directions for food, farming, pastoral and fisheries systems determined by local producers. Food sovereignty prioritizes local and national economies and markets and empowers peasant and family farmer-driven agriculture, artisanal—fishing, pastoralist-led grazing, and food production, distribution and consumption based on environmental, social and economic sustainability.[13]

The right to healthy, nutritious food, in other words, is food sovereignty, whereas the practice of growing healthy, nutritious food is agroecology. Here is a more detailed definition:

> agroecology uses intercropping, traditional fishing and mobile pastoralism, and integrates crops, trees, livestock and fish into farming practices, and also uses manure, compost, local seeds and animal breeds. It is based on ecological principles such as building life into the soil (say by adding worms into the soil), recycling nutrients, and the dynamic management of biodiversity and energy conservation. Agroecology drastically reduces our use of industrial inputs into food production. There is no use of agrotoxics (e. g., pesticides, fungicides, or herbicides), artificial hormones, or GMOs in agroecology.[14]

This has been a response to the growing dominance of industrial agriculture. Agroecology and its meaning and practices evolved internationally, including at the meetings of the World Food Summit in 1974, 1996, 2002, and 2009. The emphasis is on sustainability and eliminating waste, and therefore agroecology depends on the knowledge of local farmers. This was a genuine shift, away from industrial agriculture.[15] Why depart from industrial agriculture? It can be summed up as follows:

> The reality is that hunger in the world today is not a question of production—there is enough food on the planet, but under our current system the hungry cannot purchase it or produce enough of it. Yet the 3 billion small-scale farmers and food producers are already the ones producing 70% of the world's food. And agroecological farming by family farmers has been demonstrated to be highly productive and sustainable. It just needs more support to spread. By contrast, industrial farming currently produces only about 30% of the food consumed globally. Expanding it as if it were the solution to the global challenges of hunger, health, climate change, environmental destruction and inequity would just worsen current problems. Besides industrial agriculture uses pesticides and herbicides that deplete the soil and harm bees, and relies on monoculture, which is vulnerable to disease, extreme heat, and extreme cold.[16]

The clearest way to put it is as follows: Food sovereignty is the people's right to healthy food, whereas agroecology is the people-centered approach to growing healthy food, with the emphasis on small plots and diversity. At the 2014 Food and Agriculture Conference on Agroecology, speakers focused on the following points:

- Agroecology/organic agriculture implies a change in the scientific paradigm, replacing the tendency to destroy nature with one that seeks to produce as closely as possible in harmony with its laws. This paradigm implies that all efforts to develop and diffuse technology should be reoriented accordingly, and education, training, research and extension institutions will have to redefine their programmes and agenda.
- On the other hand, we consider that farmers' knowledge and farmers' needs should be at the center of these efforts, and farmers' and farm workers' organizations must participate in the whole process.
- Accordingly, farmers' and NGOs' successful experiences represent an important reference on which governmental and international

organizations shall rely in the process of changing the patterns of agricultural development toward sustainability.

- Urgent attention is needed to the problems of access to, and conservation of, water resources, including the question of distribution of water between neighboring countries.
- We consider that the world's "domesticated" genetic resources are a heritage of thousands of years of farmers' efforts, and this treasure cannot be appropriated for private corporate profit. Free access to genetic resources must be guaranteed for farmers and indigenous communities whose rights over the biodiversity they have improved and conserved are being threatened.
- Agroecology continues to grow, both in science and in policies. It is an approach that will help to address the challenge of ending hunger and malnutrition in all its form.
- Farmers that have adopted agroecological systems are more resilient to climate change and will recover more rapidly when faced with impacts.[17]

As already suggested, agroecology is the way to grow food, and food sovereignty is the right to food and collective food security. According to La Via Campesina, food security is achieved through local food markets, sustainable agriculture, biodiversity, democratic participation (and a rejection of monoculture and agribusiness).[18] This philosophy is quickly becoming accepted around the world (although slowly in the U.S.) Increasingly, UN agencies—the Food and Agriculture Organization and the World Food Programme—are in synch with what the world's peasants are doing and recommending. The reason for this is clear: agroecology and food sovereignty are the best defences against climate change. An emphatic statement by the UN General Assembly clarifies that agroecology is the best way that farmers have of ensuring that their crops will withstand intense heat, droughts, long dry spells, and heavy rains. The statement praises agroecology:

> It is important that adaptation policies focus on ensuring the right to food for both present and future generations through sustainable agricultural practices. This implies moving away from industrialized agricultural practices. Agroecology is an ecological approach that integrates agricultural development with relevant ecosystems. It focuses on maintaining productive agriculture that sustains yields and optimizes the use of local resources while minimizing the negative environmental and socioeconomic impacts of modern technologies . . . Small-scale farmers and

agroecological practices play a central role in conserving crop diversity and developing varieties of plants that are adapted to a range of weather conditions, including droughts.[19]

Subsequent World Food Summits increasingly emphasized that farmers with diverse crops and little reliance on pesticides and fertilizer would be much better off. So far I have briefly plotted the trajectory of agroecology—a successful, but quiet, rebellion—and have yet to fully clarify the background. Perhaps like most important struggles, who wins and who loses depends on changing conditions. Recall from Chapter 5 that the Green Revolution was touted as the extraordinary remedy for starvation and hunger, and, indeed, for food security for the world's peoples. Because of its great success and its marginalization of peasant farmers, it is worthwhile to revisit this epoch in world agriculture.

WORLD AGRICULTURE: 1940s–1960s

As already noted, Norman Ernest Borlaug, an American agronomist, is credited with launching the Green Revolution. For his pioneering work, he received the Nobel Peace Prize, the Presidential Medal of Freedom and the Congressional Gold Medal. His work began in Mexico in the 1940s where he cross-bred wheat strains that were resistant to rust with strains that were not. Then he developed high-yielding strains of wheat and rice. These may have been high yielding, but they required huge amounts of water and fertilizer, which was expensive for most farmers.

REVOLT AGAINST THE GREEN REVOLUTION

Also as briefly noted in Chapter 5, biologist Rachel Carson, especially in her 1962 book *Silent Spring*,[20] drew public attention to the dangers of pesticides and fertilizers. She had considerable influence, shaping widespread public sentiment against chemicals in agriculture, and, retrospectively, one could say that Carson was the precursor for the peasant revolution that would follow in the 1990s.

Currently, European farmers and the European Commission are on board with agroecology and seed sovereignty, although their American counterparts are not, although several U.S. groups that collaborate with Latin American farmers and do use this language, including Partners for the Land & Agricultural Needs of Traditional Peoples, or PLANT),[21] and

Women Leaders in Food and Agriculture[22] stress the same philosophy. A notable group in the U.S. is the U.S. Food Sovereignty Alliance. They describe food sovereignty as "the right of peoples to healthy and culturally appropriate food produced through ecologically sound and sustainable methods."[23]

And this from an annual report:

> Essential to life itself, seeds are complex, botanical gems providing a splendid array of food, fiber, and shelter. Through natural and human transport, seeds have migrated and adapted to rural and urban regions worldwide. Yet a handful of multinational corporations (MNCs) are intent on controlling the global seed supply, restricting producers to planting only the MNCs' patented and genetically-modified varieties, which require strict contracts and a bevy of chemical herbicides, pesticides and fertilizers. Farmers worldwide are facing the loss of seeds that their families and communities have planted for hundreds or even thousands of years.[24]

AGROECOLOGY, FOOD SOVEREIGNTY AND CLIMATE CHANGE

In 2016, the U.S. National Academies of Sciences, Engineering and Medicine, along with two other U.S. scientific bodies issued a report entitled "Genetically Engineered Crops: Experiences and Prospects."[25] These prestigious scientific organizations concluded that genetically engineering technology is "not an unqualified success," but did not conclude that it should come under scrutiny, and certainly did not endorse agroecology. This is in contrast with government and scientists' interest and support of agroecology in Canada,[26] Latin America,[27] Europe,[28] Asia and the Pacific,[29] and Africa.[30] To be sure, we are looking at a narrow window of time and evaluation of agroecology is relatively new, but it is still baffling that agroecology has been tried, praised and adopted throughout the world but not the U.S. Yet it should be noted that there are at least potential links between agroecology and the Slow Food Movement (which supports healthy, quality food).[31]

For the most part, farmers all over the world have concluded that the methods of agroecology—notably planting diverse crops, intercropping, water regulation, and having shady trees to provide cover for plants— enable plants to withstand extremes of temperature and periods of drought and rain. In other words, the peasants were right all along to be suspicious of fertilizers, chemicals, and monocropping. I am not the first to note the irony of the fact that the techniques and skills of local farmers are superior when it comes to knowing the best seeds and best practices. They have experience with the techniques of how to grow, what to grow, what not to grow in their fields, and why. As it turns out, can scientists learn from them how to deal with climate change.[32]

CONCLUSION

Feeding all the world's peoples as the population increases will be a huge challenge, and the only way it will work is through the following: promoting local food, preserving and advancing biological diversity, engaging farmer expertise and experience especially when developing new strains of plants, focusing on food crops that are healthy and resistant to climate change, and fostering "smart" fishing and agriculture practices. Alliances between scientists and farmers will greatly advance an understanding of how to grow, propagate and protect healthy food for the consumption of the world's peoples. What happened in the late 20th century was the dominance of agro-corporations and as well as the increasing obsequiousness of scientists so that the agro-corporations were in control.

Food became a commodity and the result was that agriculture was taken out of the hands of experts—namely, farmers who have intimate knowledge of soil, seeds, crops, seasons, and weather. However, crises —hunger, high rates of obesity, famines and the uncontrollable destructiveness of weedkillers—have led to the reevaluation of industrial agriculture. Indeed, there has been growing understanding of how warming increased droughts and devastated agriculture, especially mono-cropped fields. Turning to farmers and enlisting them as partners has become the solution, and, ironically, the agricultural revolution of the 21st century has meant honoring the wisdom of peasant farmers. They alone know about the effects that heavy rain, drought, storms, or intense heat can be on crops, how to save seeds from season to season, how to nurture local food culture, and to end the dependence on fossil fuels in their communities.

Accompanying the world summit to draft and approve the Paris Agreement (COP21) was the Koronivia Joint Work on Agriculture,[33] which convened again at COP23 in Bonn. Its basic goals are to ensure that farming practices are sustainable and resilient, do not contribute to climate change, and improve the well-being of farmers with smallholdings. The U.S. is not, thanks to Trump, included in this initiative.

NOTES

1 Community Alliance for Global Justice. Convention on Biological Diversity: What is it and why do we participate? Available at: https://cagj.org/2016/12/convention-on-biological-diversity-what-is-it-and-why-do-we-participate/

2 A/HRC/16/49. Report submitted by the Special Rapporteur on the Right to Food. Human Rights Council, December 17, 2010. General Assembly. Available at: www2.ohchr.org/english/issues/food/docs/A-HRC-16-49.pdf

3 Food and Agricultural Organization of the United Nations. Agroeccology knowledge hub. Available at: www.fao.org/agroecology/overview/en/

4 John Madeley, "Agroecology." *Appropriate Technology*, 43(2), 2016. Retrieved from: http://libproxy.lib.unc.edu/login?url=http://search.proquest.com/docview/1827596177?accountid=14244

5 Slow Food: The central role of food. Congress Paper 2012–2016. Available at: http://slowfood.com/filemanager/official_docs/SFCONGRESS2012__Central_role_of_food.pdf

6 US Department of Agriculture. Crops. Available at: www.nal.usda.gov/afsic/crops-0

7 La Via Campesina. Available at: www.viacampesina.org/en/index.php/organisation-mainmenu-44

8 See: La Via Campesina, ibid. Peter Andrée, Jeffrey Ayres, Michael J. Bosia, and Marie-Joséee Massicotte (Eds.) *Globalization and Food Sovereignty: Global and Local Change in the New Politics of Food* (Toronto: University of Toronto Press, 2014); and Vandana Shiva (Ed.) *Seed Sovereignty, Food Security* (Berkeley, CA: North Atlantic Books, 2016)

9 Food and Agriculture Organization of the United Nations, Weighing the GMO Arguments. Available at: www.fao.org/english/newsroom/focus/2003/gmo8.htm

10 Food and Agriculture Organization. World Food Summit. Available at: www.fao.org/docrep/x0262e/x0262e06.htm

11 Constituteproject. Available at: www.constituteproject.org/search?lang=en&q=Mother%20earth%27

12 Grace Communications Foundation. Industrial livestock production. Available at: www.sustainabletable.org/859/industrial-livestock-production

13 La Via Campesina. Declaration of Nyéléni, February 27, 2007. Available at: https://viacampesina.org/en/index.php/main-issues-mainmenu-27/food-sovereignty-and-trade-mainmenu-38/262-declaration-of-nyi

14 Via Campesina. Sustainable peasant's agriculture, March 4, 2015. Available at: https://viacampesina.org/en/index.php/main-issues-mainmenu-27/sustainable-peasants-agriculture-mainmenu-42/1749-declaration-of-the-international-forum-for-agroecology

15 Food and Agriculture Organization of the UN. World Summit on Food Security, Rome, November 16–18, 2009. Available at: www.fao.org/wsfs/world-summit/wsfs-challenges/en/

16 Agroecological Farming. Available at: www.groundswellinternational.org/how-we-work/agroecological-farming/. Stephen R. Glessman, *Agroecology: The Ecology of Sustainable Food Systems* (3rd ed.). Boca Raton: Taylor & Francis, 2015. Third World Network. *Agroecology: Key Concepts, Principles, and Practice.* Berkeley, CA: Third World Network, 2015. Available at: https://agroeco.org/wp-content/uploads/2015/11/Agroecology-training-manual-TWN-SOCLA.pdf

17   Available at: www.fao.org/about/meetings/afns/en/; http://www.fao.org/3/a-i4327e.pdf. It is sometimes recognized nevertheless that poor countries may have to go through a stage of industrial agriculture initially, since industrial farming will advance industrialization, and to have food sovereignty and agroecology requires a certain minimum level of economic security, although this is somewhat controversial. Akinwumi Adesina, May 12, 2017: Using agriculture and agribusiness to bring about industrialization in Africa. Available at: www.ipsnews.net/2017/05/using-agriculture-and-agribusiness-to-bring-about-industrialisation-in-africa/

18   Peter André, Jeffrey Avres, Michael L. Bosia, and Marie-Josée Massicotte (Eds.) *Globalization and Food Sovereignty: Global and Local Change in the New Politics of Food.* Toronto: University of Toronto Press, 2014. Vandana Shiva (Ed.) *Seed Sovereignty, Food Security.* Berkeley, CA: North Atlantic Books, 2016.

19   UN General Assembly, August 5, 2015. Right to food. Available at: www.ohchr. org/Documents/Issues/Food/A-70-287.pdf

20   Rachel Carson, *Silent Spring*. Boston, MA: Houghton Mifflin, 1962.

21   Available at: www.plantpartners.org/about-plant.html

22   League of Women in Food. Available at: www.leagueofwomeninfood.org/

23   U.S. Food Sovereignty Alliance. Food sovereignty. Available at: http://usfood sovereigntyalliance.org/what-is-food-sovereignty/

24   U.S. Food Sovereignty Alliance. *Turn the Tables on the Food Crisis*. Available at: http://usfoodsovereigntyalliance.org/wp-content/uploads/2014/04/SeedReport 2014_big.pdf

25   National Academies of Sciences, Engineering and Medicine; Division on Earth and Life Studies; Board of Agriculture and Natural Resources; Committee on Genetically Engineered Crops. Past experience and future prospects. Available at: www.nap. edu/catalog/23395/genetically-engineered-crops-experiences-and-prospects. DOI: https://doi.org/10.17226/23395

26   Food Secure Canada. Agroecology and the right to food. Available at: https://foodsecurecanada.org/resources-news/resources-research/agroecology-and-right-food

27   Miguel A. Altieri and Victor Manuel Toledo, "The agroecological revolution in Latin America." *Journal of Peasant Studies*, 38(3): 587–612: Available at: www.fao. org/family-farming/detail/en/c/386160/

28   European Commission. Agroecology in food and farming systems. Available at: https://ec.europa.eu/eip/agriculture/en/news/agroecology-food-and-farming-systems

29   Leisa India, Agroecology in Asia and the Pacific. Available at: http://leisaindia.org/articles/agroecology-in-asia-and-the-pacific/

30   Changing Course in Global Agriculture. African Agroecology Symposium, 2015. Available at: http://changingcourse-agriculture.com/a-step-forward-for-agroecology-in-africa-2/

31   Slow Food. Available at: www.slowfood.com/

32   See "Food first primer: Agroecology." Available at: https://foodfirst.org/wp-content/uploads/2014/04/FF_primer_Agroecology_Final.pdf

33   Mongabay. "Consensus grows: climate-smart agriculture key to Paris Agreement goals." Available at: https://news.mongabay.com/2017/12/consensus-grows-climate-smart-agriculture-key-to-paris-agreement-goals/

CHAPTER 7

How Bad Can It Get?

Heating up of the planet will wreck havoc with people's health and well-being, with the land, the seas, and with all living creatures. It is also important to stress that the actions and policies of one country will affect the well-being of people living in the entire world, especially if that country is large and emits huge quantities of CO_2, as does the U.S. This is precisely the problem we face today, since by not supporting the Paris Agreement, the United States will harm all peoples on the planet and the entire global environment. Therefore, I begin this chapter by highlighting that the U.S. President has potentially caused egregious harms by withdrawing from the Paris Agreement.

CRIMES AGAINST HUMANITY

This is the right time to pause to consider that Donald J. Trump can be charged for having committed a Crime against Humanity and be imprisoned for life. So long, of course, that he is consistent and affirms the decision to withdraw after November 4, 2020, which is four years after the Agreement comes into effect in the United States. Yes, his decision, if upheld, has unbelievably serious consequences, not just for some people but all people and for the whole world. Perhaps it is useful to mention how this crime is defined and applied and when the International Criminal Court has sentenced someone for it. Crimes against humanity are inhumane acts—which would constitute crimes in most of the world's national criminal law systems—committed as part of a widespread or systematic attack against civilians. This connection to a broader or systematic attack is what justifies the exercise of international criminal jurisdiction.[1] So far, 28 individuals have been convicted of Crimes against Humanity by the ICC. To give an example, Jean Leguay, a Frenchman, was convicted of this crime in 1979 for his role in an organization that arrested more than 13,000 Jews in July 16 and 17, 1942 in Paris. What will be the global consequences of the U.S. withdrawing from the Paris Agreement? Scientists are not certain, since there are so many variables, but they can make good estimates.

The science journal *Climatic Change* made the following statement:

> On average in the MENA (Middle East-North Africa), the maximum temperature during the hottest days in the recent past was about 43°C (109.4°F), which could increase to about 46°C (114.8°F) by the middle of the century and reach almost 50°C (122.0°F) by the end of the century.[2]

An article in the *National Geographic* stated the following:

> Without major reductions in emissions of greenhouse gases, such as CO_2, up to three in four people will face the threat of dying from heat by 2100. However, even with reductions, one in two people at the end of the century will likely face at least 20 days when extreme heat can kill.[3]

A further article added:

> If all the ice on land has melted and drained into the sea, raising it 216 feet and creating new shorelines for our continents and inland seas.[4]

A leading scientist who has devoted his career to studying the Arctic writes that the Arctic ice will be gone in a year or two and then will accelerate sea rise:

> the air over Greenland will get warmer and more and more of its ice will melt. It is already losing about 300 cubic kilometers of ice a year. Antarctica is adding to the melt as well. Sea-level rises will accelerate as a result.[5]

Not only scientists, but also world leaders were furious with Trump for the harm the U.S. could cause the world.

> President Macron said: "So I have two messages, one for the French researchers and scientists, we will reinforce our budgets,

our public and private investment to do more and accelerate our initiative in order to be in line with COP21."

He added:

My second message is for you guys (Americans), please to come to France, you are welcome. It is your nation, we want innovative people. We want people working on climate change, energy, renewables and new technologies. Make our planet great again.[6]

Other national leaders, including those from Japan, South Africa, Germany, Columbia, Mexico, and others, added their voices to that of Emmanuel Macron.

To be sure, this withdrawal is a process and can only be finalized in 2020. However, domestic programs related to climate change and the environment have been already been cut or eliminated by presidential orders or on the authority of Scott Pruitt, Head of the Environmental Protection Agency. The *New York Times* has a list of 52 orders that the president is abolishing or has abolished. The following is a shortlist of some of the major ones.[7]

- Slashed funding for the Environmental Protection Agency, for the Department of Energy, as well as for the climate preparedness funding to the National Oceanic and Atmospheric Administration.
- Set into motion actions to demolish the Clean Power Plan, designed to substantially reduce emissions.
- Revoked a rule that requires projects to meet higher flood requirements, a rule that coastal cities had especially welcomed.
- Repealed a ban on offshore oil and gas drilling.
- Revoked a 2016 order protecting the northern Bering Sea region (Alaska).
- Relaxed environmental review process for federal projects.
- Announced intent to stop payments to the international Green Climate Fund.
- Reopened a review of fuel efficiency standards.
- Lifted a freeze on new coal leases on public lands.
- Cancelled a requirement for oil and coal companies to report methane emissions.
- Approved the Keystone XL and Dakota Access pipelines.

The list published by the *New York Times* appeared in early October 2017. Trump (and Pruitt) continued over the next months to attack science, scientists, and the phenomena of climate change. The web pages of the Environmental Protection Agency that mentioned "climate change" were removed; laws regulating air pollution were weakened; and various steps were taken to allow oil companies to drill offshore; and steps were taken to abolish the Clean Power Plan (which had earlier been enacted to reduce emissions from power plants). Scientists were told they could not attend science meetings or present their research at meetings. For example, many scientists from the U.S. Geological Survey were told that they could not attend meetings of the American Geophysical Union or deliver research papers at the meetings.[8] In January, 2018 Trump announced that solar cells and panels from overseas would be subject to a 30% tariff, but it was widely reported that the tariffs could drive increases in pollution and endanger jobs in America's $29 billion solar industry. Then, in Trump's 2018 budget he made drastic cuts to science programs and federal science projects.

The list goes on and on, but there are two powerful countervailing forces. First, the renewable energy industry—namely, the sector of the U.S. economy responsible for the sale and installation of solar panels and wind turbines, whether imported or not—has taken off. For example, the installation of solar panels in 2016 was nearly double that in 2015. Likewise, wind power has grown, by doubling its contribution to produced energy since 2010. In other words, Donald Trump is ignoring significant facts about the American economy. Second, since U.S. cities have considerable independence from federal authority, many cities are simply ignoring Trump's rollback of federal laws and programs, and adopting plans consistent with international guidelines, and to protect their city's residents from rising water, including building walls, raising buildings, and by cutting emissions. New York City, Houston, Miami and San Francisco have all taken steps to mitigate the risks associated with rising sea levels and global temperatures.[9]

At the 2017 Conference of Mayors, over 200 U.S. mayors pledged to take action consistent with the Paris Agreement, to rely entirely on renewable energy by 2035, and many of these joined the Global Covenant of Mayors for Climate and Energy, which includes nearly 7,500 cities. The goals of the Global Covenant include setting municipal standards, which mirror those of the Paris Agreement, sharing ideas and strategies, and holding conferences.[10] The Global Covenant is more than an agreement; indeed, it is a global network, which includes the Africa Climate Change Adaptation Initiative (ACCAI) Network,[11] the Asian Cities Climate Change Resilience Network (ACCCRN)[12] and Latin American cities.[13]

In this chapter I review endangered "hotspots," some of which we may not realize are endangered by climate change because we take them so much for granted, or the endangerment process has been slow. First, I note again that scientists are now virtually certain that a 2°C degree warming by 2100 is unacceptable, and there must—there has to be—an achievable goal of 1.5°C degree warming. It is important to keep in mind that 1.5°C is achievable only if all countries pull together. It will be really hard.[14] By "scientists" here, I refer to virtually all the world's climate scientists. In this chapter I will discuss oceans, among other topics.

However, first I report the results of a study that shocked even the most cautious climate scientists. The lay public may have first learned about it in the October 18, 2017 issue of *The Guardian*.[15] The article describes "an ecological Armageddon" and reports on research by a team of German researchers published in the journal *PLOS ONE* that there will be a dramatic decline in insects over the next 20 years.[16] The reasons are complex but they all have to do with climate change. The authors also report that over the past 27 years, 75% of insects became extinct. As one scientist, quoted in *The Guardian*, remarked, "Yes, I think they are right. Come to think of it I haven't seen as many squashed bugs on my windshield."[17] The cause is industrialized farming, and specifically pesticides. The fact is that we need the bugs for pollination, for breaking down waste and dead creatures. They are the sole source of food for many amphibians, reptiles, birds, and mammals.

WHAT ELSE IS AT STAKE?

Keep in mind that scientists are cautious and do not speculate beyond what they can empirically know. I review here their greatest concerns— namely, those about which they are most worried and also most certain.

THE OCEANS

The United Nations Ocean Conference was held in New York City in June 2017.[18] The conference highlighted the sheer magnitude of the world's oceans, covering three-quarters of the earth's surface, and also highlighted that the oceans are already, and profoundly, affected by climate warming. Moreover, the dangers will increase—with a rise in ocean temperatures, ocean and coastal acidification, deoxygenation, sea-level rise, the decrease in polar ice coverage, coastal erosion and extreme weather events.[19] The speakers at the conference stressed the extraordinary contributions that the oceans make, such as being a source of food and nutrition, and yet they will have adverse effects, specifically with sea rise and as the chemistry of the oceans changes (specifically involving levels of carbon dioxide).

A subtle, but dangerous, phenomenon is that the oceans are absorbing less and less carbon dioxide as ocean waters warm. The diminished capacity of oceans to absorb CO_2 can be catastrophic since that would mean they would release more and more CO_2 into the atmosphere, dramatically increasing global warming. At this time, the oceans continue to take up about 26% of all CO_2, but if any more CO_2 would be absorbed, the oceans' acidity levels would increase and harm marine animals, fish, coral, and, indeed, all that live in the seas. In other words, it's a Catch-22 situation. Releasing CO_2 speeds up climate warming, but absorbing it is detrimental to what lives in the oceans.

To be clear, changes in seawater chemistry are occurring in oceans throughout the world. Since the beginning of the Industrial Revolution, the release of carbon dioxide from industrial and agricultural activities has increased the amount of CO_2 in the atmosphere. The ocean absorbs about a third of the CO_2 we release into the atmosphere every year, so as atmospheric CO_2 levels increase, so do the levels in the ocean. This includes CO_2 from fossil fuel burning, cement manufacture, and land-use changes. This is what is absorbed by the oceans. The especially bad news is that the oceans can absorb less CO_2 the warmer they get, which is to say that there inevitably will be increasing amounts of CO_2 in the atmosphere.[20]

A major factor in determining the rate of absorption of CO_2 by the oceans is the pace at which global CO_2 emissions are increasing over time. Over the past decades, fossil emissions (measured as tons of carbon) have increased at 2–4% annually, from around 2 billion tons in 1950 to 9 billion tons today. The oceans as a whole have a great capacity for absorbing CO_2, but ocean mixing is too slow to have spread this additional CO_2 deep into the ocean. This understanding is based on the important work of Charles David Keeling in the 1980s as a complement to his measurements of atmospheric CO_2.[21]

Studies have shown that the more acidic the environment, the more highly negative effect it has on some calcifying species, including oysters, clams, sea urchins, shallow water corals, deep sea corals, and calcareous plankton, and this is occurring throughout the world. As a report of the National Oceanic and Atmospheric Administration explains:

> When carbon dioxide (CO_2) is absorbed by seawater, chemical reactions occur that reduce seawater pH, carbonate ion

concentration, and saturation states of biologically important calcium carbonate minerals. These chemical reactions are termed "ocean acidification" or "OA" for short. Calcium carbonate minerals are the building blocks for the skeletons and shells of many marine organisms. In areas where most life now congregates in the ocean, the seawater is supersaturated with respect to calcium carbonate minerals. This means there are abundant building blocks for calcifying organisms to build their skeletons and shells. However, continued ocean acidification is causing many parts of the ocean to become undersaturated with these minerals, which is likely to affect the ability of some organisms to produce and maintain their shells."[22] This has had increasingly disastrous effects on shellfish, which are the major industry for many coastal communities worldwide.

The chemistry involved in the acidification of coral reefs is the same as it is for shellfish. Yes, the chemistry is the same but the consequences are very different because the coral reefs provide food and sanctuary for 25% of the world's marine life, and are what has been described as the "tropical forests of the oceans"[23] and as the "big cities of the sea."[24]

The bleaching of the Great Barrier Reef increases each year, and scientists report that the "2016 and 2017 bleaching events are unprecedented" and give damaged coral little chance to recover.[25] Indeed, given the importance of coral reefs and of the Great Barrier Reef in particular, the consequences are huge. The reefs not only protect fish and other marine life, but are a source of chemicals now used in medicines; they also provide nitrogen and other essential nutrients for marine food chains, assist in carbon and nitrogen fixing, protect coastlines from the damaging effects of wave action and tropical storms, help with nutrient recycling, purification of water and air, and the break-down of pollutants.[26]

Again, to clarify, the ocean's oxygen content has decreased by more than 2% over the past 50 years. That doesn't sound like much, but it is because it is particularly harmful to large fish that cannot survive in water with little oxygen. There is another worrisome trend. The amount of carbon dioxide that the oceans can hold depends on the temperature of the waters. Colder waters can absorb more carbon; warmer waters can absorb less. So, the prevailing scientific view is that as the oceans warm, they will become less and less capable of taking up carbon dioxide.[27]

THE OCEANS AND SEA RISE

Over one-third of the total human population lives within 100 km (60 miles) of an oceanic coast.[28] They will be at risk. One scenario is that the Greenland ice will melt before the end of the century, in which case the sea will rise about 23 feet. Another is that these glaciers will not melt by 2100 and, therefore, the sea will probably only rise slowly and steadily to perhaps about 2 feet. Yet even those with the most optimistic predictions increasingly caution that the Greenland glacier is melting faster than previously predicted. This not only means flooding, but a recent study highlights that an influx of freshwater from the melting ice sheet could disrupt a major monsoon system in West Africa, which in turn could dry out Africa's Sahel, a narrow region of land stretching from Mauritania in the west to Sudan in the east. The consequence could be devastating agricultural losses as the area's climate shifts. And in the most severe scenario, tens of millions of people could be forced to migrate from the area.[29]

Seas are not rising at a constant rate, but scientists find the rate at which they are rising has increased, in large part because the Greenland ice sheet is melting at a faster and faster rate.[30] This is expected to aggravate waves and storm surges.[31] Since measurements were first made in the U.S. in 1880, ocean heat, sea surface temperature and sea level have all increased, and in comparisons of 2010–2015 with 1950–1959, it is found that there has been an increase in flooding along U.S. coasts. The rising sea is threatening the very existence of small islands, and inhabitants fear they will disappear entirely. Thirty-seven small island states have formed an association for cooperation and advocacy—the Alliance of Small Island States, or AOSIS.[32] (This is discussed in greater detail in Chapter 9.) These Small Island States include Fiji that played such an important role in the COP23 conference. It is also the case that communities in the U.S.—such as Louisiana's Isle de Jean Charles and Tangier Island, Virginia, which is in the southern Chesapeake Bay—are being abandoned as they are overtaken by the sea.

CLIMATE REFUGEES

Climate change means, of course, that the earth is getting hotter. It will be insufferably hot for people living near the Equator, and it is expected that millions of people will flee their homes and become climate refugees. Really, the term is not "will" but "are." There are already climate refugees, about a million a year, and the United Nations is now coordinating a response.[33] It is important to recognize that the definition of "climate refugee" may not be broad enough to include, for example, people who have lost resources—notably land on which they grow crops or raise animals—and still have a home. It is certainly the case that Syrians fled from attacks by the forces of Bashar Hafez al-Assad and ISIS because of brutal attacks. Yet another reason is climate change—specifically, wells in Syria went dry, making it impossible to farm.[34]

It is useful to provide some comparisons. During the summer of 2017, temperatures in southern Europe soared to over 40°C (104°F), accompanied by deaths and hospitalizations. That same summer, the daytime temperature in Kuwait was 45°C (113°F).[35] Parts of northern Africa and the Middle East may be uninhabitable by 2050, according to research published in the journal *Climatic Change*.[36] Doha, Abu Dhabi, and Bandar Abbas (on the southern coast of Iran) could have temperatures of around 76°C (170°F) by 2100, but even away from the Equator it may be intolerably hot. In summer, 55°C (130°F) may be the normal daytime temperature in Phoenix.[37] Before then, we must think about alternative housing, decent accommodation, building new schools, and new workplaces.

HURRICANES

Climate change plays a major role in the formation and strength of hurricanes: First, because of sea rise, hurricanes have more water to grab hold of and then to dump. Second, a warmer ocean surface means more moisture in the atmosphere—that is, there is an increase of roughly 7% more moisture in the air for each degree Celsius of increase in sea surface temperature (SST),[38] and because warm air holds more moisture, as the atmosphere warms, rainfalls that accompany hurricanes are more likely to come in heavier downpours.

Hurricanes and cyclones that hit somewhere on the Atlantic coasts in 2017 were: Cindy (June, Louisiana), Franklin (August, Mexico City), Gert (August, Nova Scotia), Harvey (August, Texas), Irma (August–September, Barbuda and Virgin Islands), Jose (Leeward Islands, September), Katia (September, Mexico), and Maria (September, Puerto Rico and the Virgin Islands).

THE POLES: ARCTIC AND ANTARCTIC

In July 2017, a 2,200 square mile ice shelf of the Antarctic, known as Larsen C, broke free and is now adrift in the Southern Ocean. However, the ice shelf was already floating, so when it broke off and eventually will melt, it won't itself contribute much to sea-level rise, although the increasing warming of the Antarctic is extremely worrisome. Greenland is an entirely different story. Estimates are that about 270 billion tons of Greenland melt each year.[39] If the entire Greenland ice sheet were to melt, it would cause a global sea-level rise of more than 20 feet. The melting of an ice sheet like that atop Greenland can occur from the surface as air temperatures and sunlight warm the upper layer of ice. It can also occur from the edges as ice shelves collapse and fall into the oceans in large chunks.

Greenland and Larsen C are famous worldwide, but there are hundreds, if not thousands of ice glaciers. One in Peru—Mount Pucaranra—broke up in May 2017, hurling ice into a lake and destroying valuable infrastructure and threatening towns.[40] Glaciers in the U.S. are also melting fast. When President Taft created Glacier National Park in 1910, there were an estimated 150 glaciers. Since then, the number has decreased to fewer than 30, and most of those remaining have shrunk in area by two-thirds. It is predicted that within 30 years most, if not all of the park's glaciers will disappear. There are about 198,000 glaciers in the world, covering 726 million square miles, and if they all melted, they would raise sea levels by over a foot, or about 405 mm.[41]

Also, higher spring and summer temperatures and earlier snow melt cause soils to be drier for longer, increasing the likelihood of drought and a longer wildfire season, particularly in the western United States. The wildfires north of San Francisco in October 2017 were deadly and the fiercest ever recorded in California's history. Yet these were followed by fierce fires in California in December. Warming plays a role because it dries out vegetation—namely, aggravating the condition for more fires, and as areas burn, carbon is released, aggravating the conditions for climate change.

WEATHER, AIR, AND HEALTH

The Union of Concerned Scientists (UCS) reports that as the air warms (which is inevitable with climate change), moisture and precipitation levels will change. Rainfall reduction and soaring temperatures have already taken their toll on African agriculture, through which the majority of people earn their living. Wet areas will become wetter and dry areas will become drier. It is important, too, that air pollution will get worse.

People's health is of great concern. Research reported in the *Lancet* estimates that two-thirds of the European population will be affected annually by weather-related disasters, such as heatwaves, posing a risk to health and lives.[42] The World Health Organization makes some dire predictions about climate change and how it will affect the social and environmental determinants of health, including clean air, safe drinking water, food sufficiency, and secure shelter. It estimates that between 2030 and 2050, climate change can be expected to cause approximately 250,000 additional deaths per year, from malnutrition, malaria, diarrhea and heat stress, and that the direct damage costs to health—i.e. excluding costs in health-determining sectors such as agriculture and water and sanitation—is estimated to be between US\$2–4 billion/year by 2030.[43] It was sadly confirmed at the September 2017 International Conference on Nutrition (ICN), that even the most sturdy crops—varieties of grain—lose their nutritional value each year.[44] This is disastrous since the peoples in the poorest parts of the world—Southeast Asia and Africa—depend on grain as a source of iron and protein.

Air pollution and climate change are closely related. The main sources of CO_2 emissions—the extraction and burning of fossil fuels—drive climate change, but are also major sources of air pollutants. Furthermore, many air pollutants that are harmful to human health and the environment also contribute to climate change because they affect how much sunlight is reflected or absorbed by the atmosphere, with some pollutants warming and others cooling the earth. These so-called short-lived climate-forcing pollutants include methane, black carbon (soot), ground-level ozone, and some aerosols. They have significant impacts on the climate; carbon and methane in particular are among the top contributors to global warming after CO_2.[45]

The strictest laws (Protocol) governing pollution have been set by the United Nations Economic Commission for Europe (UNECE). The Protocol sets national emission ceilings for 2010 up to 2020 for four pollutants: sulfur, nitrogen oxides, volatile organic compounds (VOCs), and ammonia.[46] The Protocol also sets tight limit values for specific emission sources (e.g. combustion plant, electricity production, dry cleaning, cars, and trucks) and requires best available techniques to be used to keep emissions down. VOCs emissions from such products as paints or aerosols also have to be cut. Finally, farmers have to take specific measures to control ammonia emissions.

PLANTS AND ANIMALS

Global warming impacts the migration of plants and animals in the following ways: 1) forced migration of plants and animals to higher altitudes, with the possibility of extinction; 2) increase in agricultural pests drives migration; 3) desynchronization of life-cycle events—for example, in bird migration; 4) changing woodlands—for example, when tree species attempt to migrate; and increase in allergens and noxious plants, such as migration of poison ivy.[47] Certainly, forestry and agriculture will be affected with warmer air and changing conditions of precipitation. It is also well known that as the temperature of the oceans rise, corals die and fish migrate to warmer waters.

A study found that more than 450 plants and animals have undergone local extinctions due to climate change. It is useful to consider the list of the World Wildlife Fund.[48] A few on its list of 66 vulnerable, endangered, and critically endangered species are the Adélie Land Penguin, Armur Leopard, Black Rhino, Bornean Orangutan, Cross River Gorilla, Eastern Lowland Gorilla, Hawksbill Turtle, Javan Rhino, Malayan Tiger, Mountain Gorilla, Orangutan, South China Tiger, Sumatran Elephant, Sumatran Orangutan, Sumatran Rhino, Sumatran Tiger, Western Lowland Gorilla, Yangtze Finless Porpoise, and the Snow Leopard. Of gravest concern are coral, on which millions of people depend for their livelihoods, and which provide home for a quarter of all fish species. Still, this is just the tip of the iceberg. Scientists now expect one in six species to be extinct by the end of the century,[49] and, when taken along with the scientific prediction about the dramatic loss of instincts (the "ecological Armageddon"), one can only conclude that it is imperative to limit warming to 1.5°C.

CONCLUSIONS

No continent, no country, city, town, or person will be unaffected by the devastating effects of the climate as it is turning to be an enemy of human beings. The whole idea of the Paris Agreement is solidarity and global cooperation to protect us all from heating of the planet. Some countries will be uninhabitable, and that means that the others must accept refugees. Some grains will become extinct, and that means we must breed new sturdier varieties. Hurricanes and wildfires will destroy entire communities and that means we must prepare by building better structures. Oceans will continue to rise and that means we will need to build walls and create seaworthy housing.

Trump's decision that the U.S. leave the Paris Agreement will not only fatefully harm all Americans, it will harm the entire world. In this chapter, I have captured some of the consequences based on scientists' conclusions, but not even scientists can predict the full consequences of the U.S. leaving Paris. There are many unknowns: Will storms increase in intensity everywhere? How soon will the Arctic melt? When will small island states be overtaken by the sea? Will countries develop humane and welcoming policies for climate refugees? Will the U.S. remain in Paris, with Trump graciously rescinding his decision?

NOTES

1 International Crime Database. Available at: www.internationalcrimesdatabase.org/Crimes/CrimesAgainstHumanity#p3

2 J. Lelieveld, Y. Proestos, P. Hadjinicolaou, M. Tanarhte, E. Tyrlis, G. Zittis. "Strongly increasing heat extremes in the Middle East and North Africa (MENA) in the 21st century." *Climatic Change*. July 2016, 137(1–2): 245–260. Available at: https://link.springer.com/article/10.1007%2Fs10584-016-1665-6

3 *National Geographic*. "By 2100, deadly heat may yhreaten majority of humankind." Available at: https://news.nationalgeographic.com/2017/06/heatwaves-climate-change-global-warming/

4 *National Geographic*. "What the world would look like if all the ice melted." Available at: www.nationalgeographic.com/magazine/2013/09/rising-seas-ice-melt-new-shoreline-maps/

5 Peter Wadhams. *Farewell to Ice*. Penguin Books, 2016.

6 Alex Ward, "French President Emmanuel Macron responds to Trump: 'Make our planet great again.' " *Vox*, June 1, 2017. Available at: www.vox.com/world/2017/6/1/15727140/emmanuel-macron-trump-paris-agreement-make-our-planet-great-again

7 *New York Times*. "Fifty-two environmental rules on the way out under Trump." October 5, 2017 (updated). Available at: www.nytimes.com/interactive/2017/10/05/climate/trump-environment-rules-reversed.html?_r=0

8 Sarah Kaplan, "Government scientists blocked from the biggest meeting in their field." *Washington Post*, December 22, 2007. Available at: www.washington post.com/news/speaking-of-science/wp/2017/12/22/government-scientists-blocked-from-the-biggest-meeting-in-their-field/?utm_term=.5d3dc029991e

9 *The Guardian*. "The fight against climate change: Four cities leading the way in the Trump era." June 12, 2017. Available at: www.theguardian.com/cities/2017/jun/12/climate-change-trump-new-york-city-san-francisco-houston-miami

10 Covenant of Mayors. Covenant of Mayors for Climate and Energy [EU]. Available at: www.covenantofmayors.eu/index_en.html

11 Africa Climate Change Adaption Initiative. Available at: www.accai.net/

12 Asian Cities Climate Change Resilience Network. Available at: www.accrn.net/

13 United Nations Framework Convention on Climate Change. "Latin American cities commit to act on climate change." Available at: http://newsroom.unfccc.int/financial-flows/latin-american-cities-commit-to-act-on-climate/

14 Adrian E. Raftery, Alec Zimmer, Dargan M.W. Frierson, Richard Startz, and Peiran Liu, "Less than 2°C warming by 2100 unlikely." *Nature Climate Change*, 7: 637–641, July 31, 2017. Available at: www.nature.com/nclimate/journal/v7/n9/full/nclimate3352.html

15 Damian Carrington, "Warning of 'ecological Armageddon' after dramatic plunge in insect numbers." *The Guardian*. Available at: www.theguardian.com/environ ment/2017/oct/18/warning-of-ecological-armageddon-after-dramatic-plunge-in-insect-numbers

16 C.A. Hallmann, et al. (2017) More than 75 percent decline over 27 years in total flying insect biomass in protected areas. *PLOS ONE*, 12(10): e0185809. Available at: https://doi.org/10.1371/journal.pone.0185809

17 Michael McCarthy, "A giant insect ecosystem is collapsing due to humans. It's a catastrophe." *The Guardian*. Available at: www.theguardian.com/environment/2017/oct/21/insects-giant-ecosystem-collapsing-human-activity-catastrophe

18 UN Ocean Conference, June, 5–9 June 2017. Available at: https://oceanconference. un.org/

19 UN Ocean Conference: Call for Action. Available at: https://oceanconference. un.org/callforaction

20 Climate Home. "Warming oceans face CO_2 tipping point." Available at: www. climatechangenews.com/2012/01/24/warming-oceans-face-co2-tipping-point/

21 Scripps Institution of Oceanography. "How much CO_2 can the oceans take up?" Available at: https://scripps.ucsd.edu/programs/keelingcurve/2013/07/03/how-much-co2-can-the-oceans-take-up/

22 PMEL NOAA. What is ocean acidification? Available at: www.pmel.noaa.gov/co2/ story/What+is+Ocean+Acidification%3F

23 World Wildlife Fund (WWF). "Coral reefs." Available at: wwf.panda.org/about_ our_earth/blue_planet/coasts/coral_reefs/

24 Smithsonian National Museum of Natural History. "Corals and coral reefs." Available at: http://ocean.si.edu/corals-and-coral-reefs

25 Christopher Knaus and Nick Evershed, "Great Barrier Reef at 'terminal stage': Scientists despair at latest coral bleaching data." Available at: www.theguardian.com/ environment/2017/apr/10/great-barrier-reef-terminal-stage-australia-scientists-despair-latest-coral-bleaching-data

26 Biodiscovery and the Great Barrier Reef. "Human impact on the reef." Available at: www.qm.qld.gov.au/microsites/biodiscovery/05human-impact/importance-of-coral-reefs.html. Daniel Cressey, "Coral crisis: Great Barrier Reef bleaching is 'the worst ever seen'." *Nature*, April 13, 2016. Available at: www.nature.com/news/coral-crisis-great-barrier-reef-bleaching-is-the-worst-we-ve-ever-seen-1.19747

27 Tim DeVries, Mark Holzer, and François Primeau, "Recent increase in oceanic carbon uptake driven by weaker upper-ocean overturning." *Nature*, February 9, 2017. Available at: www.nature.com/nature/journal/v542/n7640/full/nature21068. html. Besides, because the oceans are warming, fish (and lobsters) are swimming north. Rod Fujita, "5 ways climate change is affecting our oceans." Environmental Defense Fund, October 8, 2013. Available at: www.edf.org/blog/2013/11/14/five-ways-climate-change-affecting-our-oceans

28 NASA, "Living Ocean." Available at: https://science.nasa.gov/earth-science/ocean ography/living-ocean

29 David Defrance, "Consequences of rapid ice sheet melting on the Sahelian population vulnerability." May 11, 2017. Available at: www.pnas.org/content/ 114/25/6533

30 Sean Vitousek, Patrick L. Barnard, Charles H. Fletcher, Li Erikson, and Curt Storlazzi, "Doubling of coastal flooding frequency with decades due to sea-level

rise. *Nature*, March 23, 2017. Available at: www.nature.com/articles/s41598-017-01362-7

31   Damian Carrington, "Sea level rise will double coastal flood risk worldwide." *The Guardian*, May 18, 2017. Available at: www.theguardian.com/environment/2017/may/18/sea-level-rise-double-coastal-flood-risk-worldwide

32   Available at: http://aosis.org/

33   Available at: http://disasterdisplacement.org/

34   *Smithsonian* magazine. Is a lack of water to blame for the conflict in Syria? June, 2013. Available at: www.smithsonianmag.com/innovation/is-a-lack-of-water-to-blame-for-the-conflict-in-syria-72513729/

35   Ruth Michaelson, "Kuwait's inferno." *The Guardian*, August 18, 2017. Available at: www.theguardian.com/cities/2017/aug/18/kuwait-city-hottest-place-earth-climate-change-gulf-oil-temperatures

36   J. Lelieveld, Y. Proestos, P. Hadjinicolaou, M. Tanarhte, E. Tyrlis, and G. Zittis, "Strongly increasing heat extremes in the Middle East and North Africa (MENA) in the 21st century." *Climatic Change*, 137, 1–2: 245–260. Available at: https://link.springer.com/article/10.1007%2Fs10584-016-1665-6

37   *New York Times*. "Think it's hot now?" Available at: www.nytimes.com/interactive/2016/08/20/sunday-review/climate-change-hot-future.html?mcubz=0&_r=0

38   This is expressed by the Clausius-Clapeyron equation. Michael E. Mann, Thomas C. Peterson, and Susan Joy Hassol, "What we know about the climate change–hurricane connection." *Scientific American*, September 8, 2017. Available at: https://blogs.scientificamerican.com/observations/what-we-know-about-the-climate-change-hurricane-connection/

39   B. Noel et al. "A tipping point in refreezing accelerates mass loss of Greenland's glacier and ice caps." *Nature Communications*, 8, March 31, 2017. Available at: www.nature.com/articles/ncomms14730

40   Ben Orlove, "Palcacocha icefalls demonstrate hazard vulnerabilities in Peru." Phys.org, June 12, 2017. Available at: https://phys.org/news/2017-06-palcacocha-icefalls-hazard-vulnerabilities-peru.html#jCp

41   Bethan Davies, "Mapping the world's glaciers." Antarctic Glaciers.org, June 22, 2017. Available at: www.antarcticglaciers.org/glaciers-and-climate/glacier-recession/mapping-worlds-glaciers/

42   Giovanni Forzieri, Alessandro Cescatti, Filipe Batista e Silva, and Luc Feyen, "Increasing risk over time of weather-related hazards to the European population." *The Lancet: Planetary Health*, 1(5), August 2017. Available at: www.thelancet.com/journals/lanplh/article/PIIS2542-5196(17)30082-7/fulltext

43   World Health Organization. Climate change and health. Available at: www.who.int/mediacentre/factsheets/fs266/en/

44   International Symposium. Food security and nutrition in the age of climate change, September 24–27, 2017. Available at: www.mrif.gouv.qc.ca/en/salle-de-presse/evenements-speciaux/Colloque-san/programme

45   Institute for Advanced Sustainability Studies. Air pollution and climate change. Available at: www.iass-potsdam.de/en/content/air-pollution-and-climate-change

46   United Nations Economic Commission for Europe. Protocol to abate acidification, eutrophication and ground-level ozone. Available at: www.unece.org/env/lrtap/multi_h1.html

47 Union of Concerned Scientists. Climate hot map. Available at: www.climatehotmap.org/global-warming-effects/plants-and-animals.html
48 World Wildlife Fund. Species list. Available at: www.worldwildlife.org/species/directory
49 Available at: http://science.sciencemag.org/content/348/6234/571.full

CHAPTER 8

Rising Seas and
Tiny Countries

I begin this chapter describing small—indeed, tiny—countries: Barbuda and Antigua (and, more specifically, Barbuda), Kiribati, and Costa Rica. Although they are very different in many respects, all of them are extremely vulnerable to what accompanies climate warming—notably storms, hurricanes, and sea rise, and, yes, the destruction of homes and property, and often deaths and injuries. One—Barbuda—is in the Caribbean; another—Kiribati—is in the Central Pacific; and Costa Rica is in Central America. All share ocean environments, but they differ in their vulnerabilities to climate change. Let's start with Barbuda to understand how climate change is fueling hurricanes, and then turn to Kiribati to see how climate change is responsible for sea level rise. Then, I turn to Costa Rica as an example of a country relatively advantaged by its location, enabling its peoples to creatively deal with climate change. Finally, I contend that these three countries and their peoples are not unique. There are millions of people who are now at risk, or will be at risk, of losing their homes and their livelihoods because of the rising seas. Finally, I briefly describe the immense danger Puerto Ricans have faced (and continue to do so) and that Marshall Islanders are now encountering.

This chapter mainly highlights the grave and horrific consequences of storms whipped up by the warming oceans, but it can get worse—much worse. If Trump doesn't rejoin the Paris Agreement for the U.S., the consequences will be unimaginable for the whole world because the U.S. is huge and spews out an incredible amount of carbon dioxide—in fact, the U.S. is the second biggest emitter after China. I would, however, like here to anticipate a hopeful point I make in the final chapter, quoting the Fijian hosts of COP23: "We are all in the same canoe." That means everyone.

BARBUDA

On the September 15, 2017, *New York Post* carried the following headline: "Civilization on Barbuda has been extinguished by Irma."[1] Another headline in the British paper, *The Independent*, posed the question: "What it's like in Barbuda, the island ripped apart by Irma and forgotten by the world."[2] And CNN announced, "For the first time in 300 years, no one is living on Barbuda."[3] Yes, Barbuda was totally and completely destroyed by Irma and its 1,700 residents were all evacuated 39 miles away to Antigua, Barbuda's sister island. Nothing is left of Barbuda and it has been described as a "humanitarian disaster" and as a "ghastly apocalyptic scene." (Note that the communications were so poor and the country so isolated that these reports by the *Post*, the *Independent*, CNN and other news outlets appeared nine days after Barbuda was destroyed.)

As a brief background, Antigua and Barbuda is a sovereign commonwealth with the queen of England as its head. The original inhabitants were indigenous peoples, and when the English colonized the islands, they introduced slavery. The islands became independent in 1981. Each island has its own outlying islands; Antigua has 21 and Barbuda has 4. Every single island is stunningly beautiful—and, yes, they are extremely vulnerable. Antigua is higher (with an average elevation of a little more than 1,300 feet, or 402 meters) and Barbuda with an average elevation of 23 feet, or 5 meters above sea level. (In comparison, Central Park in New York City is 36 meters, or 118 feet, above sea level.). Given this elevation, it is understandable why Barbuda was unable to withstand Irma.

Rebuilding Barbuda is expected to cost between $250 million and $300 million, an impossible burden for a country for which the GDP is ranked 113th in the world.[4] These were estimates from Irma. Barbuda was hit again (though not hard) on September 20 by Maria. How can Barbuda possibly rebuild?[5] Through the fall and still at the beginning of December 2017 Barbuda was abandoned.

THE CONTEXT

Irma was a category four hurricane when it hit Barbuda on September 5. Earlier, on August 25, Hurricane Harvey smashed into Rockland, Texas, and continued to devastate Dallas. Harvey was also a category four hurricane. On September 9, Katia slammed into Mexico, killing two people. Katia did not last long, but it was also a category four hurricane. On September 20, Maria pounded Puerto Rico. Maria, too, was a category four hurricane. On September 21, Jose was building up energy in the Atlantic off New England. It was a tropical storm.

How can we account for this unusual activity and the effects? The explanation is climate change. Just to recapitulate, CO_2 and methane in the atmosphere block and prevent some of the sun's rays from bouncing back into space, thereby trapping heat and raising air temperatures all over the world. As the air warms, some of that heat is absorbed by the ocean, which in turn raises the temperature of the sea's upper layers, and the warmer the ocean the more likely it is there will be a hurricane and the stronger it will be. Besides, atmospheric moisture increases with warming which means more rain.[6] The large amount of moisture creates the potential for much greater rainfall and greater flooding. Not only were the surface waters of the Caribbean unusually warm in September, but there was a deep layer of warm water and that helped fuel hurricanes.

KIRIBATI

With this background, we can now turn to Kiribati. Extraordinarily beautiful, Kiribati has white sand beaches and clear ocean waters. Tourists come to fish, swim, to snorkel, dive, and surf. Kiribati (pronounced KIRR-i-bas) has three island groups—Gilbert Islands, Line Islands, and Phoenix Islands—with a total land area of 811 square kilometers. The country consists of 32 atolls and one raised coral island dispersed over 3.5 million square kilometers, straddling the Equator, and bordering the International Date Line. The terrain is mostly low-lying coral atolls surrounded by extensive reefs, with a total coastal area of 1,143 kilometers.[7] The country is dispersed over 3.5 million square kilometers, making communications, travel, and the delivery of goods and services particularly challenging. (The capital, Tarawa, is about half way between Hawaii and Australia.) The estimated population of the Republic of Kiribati in 2009 was 112,850, with 21 of the 33 islands inhabited. It was a colony of Great Britain, achieving independence in 1979, and now a member of the Commonwealth.

Things are not going at all well for Kiribati. One reason for this is that the county ranks 150th (out of 195) on GDP, a fairly good indicator of the standard of living.[8] Another is that it is becoming increasingly clear that the sea is overtaking Kiribati. In June 2008, President Anote Tong said:

> We may already be at the point of no return, where the emissions in the atmosphere will carry on contributing to climate change, so in time our small low-lying islands will be submerged. According to the worst case scenarios, Kiribati will be submerged within (this) century.

He added that climate change "is not an issue of economic development; it's an issue of human survival."[9] A UN report states: "In Kiribati, almost every household (94%) was impacted by natural hazards over the preceding 10 years, with 81% of them indicating that they had been affected by sea-level rise."[10]

Some of Kiribati's 94,000 people who lived in shoreline village communities have already been relocated. "We're doing it now. . . it's that urgent," Tong said. The government of Kiribati bought nearly 6,000 acres in Fiji, an island nation more than 1,000 miles away, as a potential refuge. Fiji's higher elevation and more stable shoreline make it less vulnerable. While the government is finding and promoting ways to advance ways the people can remain, including developing water-treatment facilities, plans are evolving for the migration to Fiji, which is accompanied by the expression, "migration with dignity."[11] To be sure, residents are themselves aware of their own risks with climate change, as many have relocated to escape sea rise and the country has played an active role in international climate meetings. Importantly, Fiji presided over the November 2017 Climate Change Conference which was held in Bonn, Germany.[12]

COSTA RICA

This Caribbean country is situated in Central America, and, as I will discuss, has not only responded to the challenges of climate change but has been an exceptional pioneer. It has the same climate that Kiribati and Barbuda have, but its location and elevation protect it from horrific storms and sea rise—that is to say, almost. There has been a notable change in the microclimates of the forests that has affected trees, plants, birds, and animals. Climate scientists have highlighted concerns for Costa Rica as it will be affected by sea rise from both the Pacific and Atlantic oceans, creating instability of the land mass. Besides, its 34 islands are at high risk from sea rise and hurricanes.

Before I proceed, I will provide a brief, introduction to its government and institutions. Costa Rico is a stable, progressive democracy, with an outstanding public education system, high participation rates of women in government, and a modern public health system. There is a multi-party system and regular elections for legislators. Costa Rica's value on the UN's Human Development Index (HDI) is .776, which puts the country in the high human development category, and, more specifically, is 66th out of 188 countries. To quote from the UN report:

Between 1990 and 2015, Costa Rica's HDI value increased from 0.653 to 0.776, an increase of 18.9 percent. Between 1990 and 2015, Costa Rica's life expectancy at birth increased by 4.0 years, mean years of schooling increased by 1.8 years and expected years of schooling increased by 4.4 years. Costa Rica's GNI per capita increased by about 98.5 percent between 1990 and 2015.[13]

Costa Rica has mountains, beaches, rainforests, volcanoes, and waterfalls. On top of the tallest mountain, Cerro Chirripo, one can see both the Caribbean and Pacific coasts. The country is exceptional with respect to biodiversity and forest management. The plan has been to increase the acreage devoted to forests, and this has been hugely successful, going from forest cover of 21% in 1983 to 52% by 2012. The point of this, of course, is to help achieve carbon neutrality. Costa Ricans are relying on the ability of these growing forests so that their goal can be achieved. According to official estimates, 75% of its carbon dioxide can be absorbed by forests in 2021. This is the year in which Costa Rica is aiming to achieve its goal and become the first carbon-neutral country in 2021. This plan is accompanied by a deadline to ban single-use plastic bags, plastic straws, and plastic wrapping by 2021.[14]

Preserving energy, preservation of the environment, carbon neutrality, and climate are all key features of the society and government. For example, climate change is part of the educational curriculum. Furthermore, Climate Change Strategy (Estrategia Nacional de Cambio Climático, ENCC) is the basis for Costa Rica's goal of achieving carbon neutrality by 2021. Its launch was supported by the creation of a Climate Change Department (DCC) at the Ministry of Environment and Energy (MINAE) in charge of implementing and following up on international commitments and implementation of policies with regard to climate change.[15] Furthermore, in the 2011 revision of the Costa Rico constitution, articles were added that deal with protection of the environment.[16]

Article 46: Consumers and users have the right to the protection of their health, environment, security and economical interests, to receive adequate and true information; to the freedom of election, and to an equitable treatment. The state will support the organs constituted by them for the defense of their rights. The law will regulate these matters.

Article 50: All persons have the right to a healthy and ecologically balanced environment. For that, they are legitimated to denounce the acts that infringe this right and to claim reparation for the damage caused.

Costa Rica is exemplary with its goal to be carbon neutral by 2021, and it appears to be able to meet this goal. In 2015, it ran on renewables for 75 days straight, relying solely on hydroelectric power.[17] Although the country has some wind turbines and sources for solar energy, it relies mainly on its 14 hydroelectric plants,[18] and in all respects has been innovative in developing climate policies that advance sustainability.[19]

INTERNATIONAL COALITIONS

In this chapter I have highlighted the devastation to Barbuda, the horrific perils faced by residents of Kiribati, and the remarkable innovativeness of Costa Rica, which may be the first country to rely solely on renewables to meet its energy needs. The mainland of Costa Rica is relatively well protected from hurricanes, but its 34 islands are not, providing a powerful incentive to be an innovative player on the world stage.

In the face of these perilous possibilities, these three countries, along with others, have organized two international alliances. One is the "Vulnerable 20," which started out with 20 members: Afghanistan, Bangladesh, Barbados, Bhutan, Costa Rica, Ethiopia, Ghana, Kenya, Kiribati, Madagascar, Maldives, Nepal, Philippines, Rwanda, Saint Lucia, Tanzania, Timor-Leste, Tuvalu, Vanuatu, and Vietnam.[20] Their main objective is to strengthen international provisions to assist poor countries finance climate projects and to promote the Green Climate Fund. They met with representatives from the World Bank in October 2015.[21] Yet by 2016, the group highlighted substantive concerns about climate change in addition to financial concerns, and attracted more members: Burkina Faso, Cambodia, Colombia, Comoros, Democratic Republic of the Congo, Dominican Republic, Fiji, the Gambia, Grenada, Guatemala, Haiti, Honduras, Lebanon, Malawi, Marshall Islands, Mongolia, Morocco, Niger, Palau, Palestine, Papua New Guinea, Senegal, South Sudan, Sri Lanka, Samoa, Sudan, Tunisia, and Yemen. Now, the enlarged group, the Climate Vulnerable Forum, has played a growing role in international action to promote renewable energy, has set deadlines for ending fossil fuels, and insists on a warming increase of no more than 1.5 Celsius. [22]

The other international alliance is made up of low-lying, small states that are at risk of being overtaken by the sea. It is the Alliance of Small Island States (AOSIS) and is made up of 36 member states: Antigua and Barbuda, Bahamas, Barbados, Belize, Cape Verde, Comoros, Cook Islands, Cuba, Dominica, Dominican Republic, Fiji, Federated States of Micronesia, Grenada, Guinea-Bissau, Guyana, Haiti, Jamaica, Kiribati, Maldives, Marshall Islands, Mauritius, Nauru, Niue, Palau, Papua New Guinea, Samoa, Singapore, Seychelles, Sao Tome and Principe, Solomon Islands, St. Kitts and Nevis, St. Lucia, St. Vincent and the Grenadines,

Suriname, Timor-Leste, Tonga, Trinidad and Tobago, Tuvalu, and Vanuatu.[23] As an alliance and as individual states, coalition members have applied for and received funds to reduce risks related to climate change—for example, to purchase wind turbines, or promote marine biodiversity—and AOSIS has protested attempts to dilute environmental and climate guidelines.[24]

The Marshall Islands is a member of AOSIS. The *New York Times* bluntly described the islands in these terms: "The Marshall Islands are Disappearing," and, in fact, that can be said for all the states in the alliance. The following is an excerpt from the article in the *New York Times*:

> Most of the 1,000 or so Marshall Islands, spread out over
> 29 narrow coral atolls in the South Pacific, are less than six feet
> above sea level—and few are more than a mile wide. For the
> Marshallese, the destructive power of the rising seas is already
> an inescapable part of daily life. Changing global trade winds
> have raised sea levels in the South Pacific about a foot over the
> past 30 years, faster than elsewhere. Scientists are studying
> whether those changing trade winds have anything to do
> with climate change. But add to this problem a future sea-level
> rise wrought by climate change, and islanders who today
> experience deluges of tidal flooding once every month or two
> could see their homes unfit for human habitation within the
> coming decades.[25]

Yet the Marshall Islands has a unique relationship to the U.S. The Marshall Islands no longer "belong" to the U.S. because it signed a Compact of Free Association with the United States in 1983 and gained independence in 1986. But under contract, the U.S. Army uses the Kwajalein Atoll missile test range for nuclear and hydrogen tests. In April 2014, the Marshall Islands filed suits against the United States and other nuclear powers claiming failure to meet their obligations under Article VI of the Nuclear Non-Proliferation Treaty. The suits do not seek any monetary compensation for the Marshall Islands-related nuclear legacy. On February 3, 2015, the federal court in California dismissed the lawsuit.[26]

Most of the thousand or so Marshall Islands, spread out over 29 narrow coral atolls in the South Pacific, are less than 6 feet above sea level—and few are more than a mile wide. For the Marshallese, the destructive power of the rising seas is already experienced, and floods are becoming more frequent and more destructive.

PUERTO RICO

Another betrayal by the U.S. government (that is, Trump) has been
Puerto Rico. Making landfall on September 20, 2017, Hurricane Maria
wreaked havoc on the island, causing a level of widespread destruction
and disorganization paralleled by few storms in American history. Making
a direct hit on the island, with 50- to 60-mile-wide reach, it lasted for 30
hours. Trump was nasty and vindictive, accusing Puerto Ricans of being
lazy, unappreciative, and whining. In January, 2018, a third of Puerto
Ricans still lacked electricity and many did not have running water, but
in early February, the Federal Emergency Management Agency (FEMA)
announced it was leaving and would no longer provide assistance. Puerto
Ricans were stunned. However, in response to thousands of signatures on
petitions and protests by Congress people, FEMA returned to help provide
services.

CONCLUSION

Small Island States have been at the forefront of climate action. At the
meetings for the Paris Agreement, they pressed for stronger action and to
include 1.5°C in the final agreement (and now scientists concur that a rise
of 2°C would be disastrous.[27]) They have been internationally active. For
example, Fiji, an active member of AOSIS, presided at the COP23
Climate Conference which was held in Bonn, in November 2017. An
initiative undertaken by members of the Climate Vulnerable Forum as
well as the Alliance of Small Island States has been to ban plastic bags or
to set deadlines to prohibit them. Yes, they have been leaders in the
climate movement and for good reason—they are most at risk. Just in
2017 alone, five of the Solomon Islands were overtaken by the sea, and
it is only a matter of time that other small islands will be as well.

Costa Rica has played a global leadership role. Although mainly
located on the mainland of Central America, the country has islands that
are at risk and perhaps for that reason has joined the Climate Vulnerable
Forum in which they have been especially active. Indeed, the country is
committed to becoming the first carbon-neutral economy by 2021. It is
useful to close this chapter with a list of the top five American cities that
are most highly vulnerable to major coastal flooding and sea rise by 2050:
New York City, Hialeah, Miami, Fort Lauderdale, and Pembroke Pines.[28]
The good news is that cities are creatively adopting responses to climate
warming that are consistent with the Paris Agreement.

Trump seems not to have understood what climate change is: that the seas are rising; that some parts of the world will be so hot that they will not be habitable; that the Arctic is quickly melting. It would be indulgent to him and harmful to the world to allow him to stay in office.

NOTES

1   *New York Post*, "Civilization has been "extinguished" by Irma." September 15, 2017. Available at: http://nypost.com/2017/09/15/civilization-on-barbuda-has-been-extinguished-by-irma/

2   *The Independent*, "What it's like in Barbuda, the island ripped apart by Irma and forgotten by the world." September 2017. Available at: www.independent.co.uk/news/world/americas/irma-barbuda-hurricane-island-damage-what-it-is-like-latest-a7941516.html

3   CNN, "Irma has left Barbuda uninhabitable." September 15, 2017. Available at: www.cnn.com/videos/us/2017/09/15/irma-left-barbuda-uninhabitable-orig-trnd-lab.cnn

4   There is a possible happy ending. In 2004 the World Trade Organization ruled that the U.S. had violated trade agreements by not allowing online betting from websites hosted in Antigua and Barbuda. An appellate body upheld the same decision the following year. So far the U.S. has not paid. NBC News. "Barbuda hopes online betting settlement can aid Irma recovery." September 20, 2017. Available at: www.nbcnews.com/news/us-news/barbuda-hopes-online-betting-settlement-can-aid-irma-recovery-n802591. And additionally, thanks to the generosity of a very rich actor, Robert De Niro, Barbuda may fare all right.

5   Prensa Latina. "Antigua and Barbuda lightly hammered by Hurricane Maria." September 25, 2017. Available at: www.plenglish.com/index.php?o=rn&id=18374&SEO=antigua-and-barbuda-lightly-hammered-by-hurricane-maria

6   See "Clausius–Clapeyron relation." *Wikipedia*. Available at: https://en.wikipedia.org/wiki/Clausius%E2%80%93Clapeyron_relation

7   Office of the President, Republic of Kiribati. "Climate change." Available at: www.climate.gov.ki/about-kiribati/

8   United Nations Statistics Division, National Accounts. Available at: https://unstats.un.org/unsd/snaama/selbasicFast.asp

9   *Daily Express*. "Sinking nation applies for help." June 9, 2008. Available at: www.express.co.uk/news/world/47565/Sinking-nation-appeals-for-help

10 United Nations Framework Convention on Climate Change. "Pacific Islanders faced with climate-induced migration can benefit from Paris Agreement." February 9, 2017. Available at: http://newsroom.unfccc.int/cop-23-bonn/pacific-islanders-faced-with-migration-can-benefit-from-paris-agreement/

11 Office of the President of Kiribati. "Relocation." Available at: www.climate.gov.ki/category/action/relocation/

12 Available at: http://unfccc.int/meetings/items/6240.php

13 United Nations Development Program. "Briefing notes for countries on the 2016 Human Development Report. Costa Rica." Available at: http://hdr.undp.org/sites/all/themes/hdr_theme/country-notes/CRI.pdf

14 *Lifegate.* "Costa Rica aims to become the first country to ban single-use plastics." September 15, 2017. Available at: www.lifegate.com/people/news/costa-rica-single-use-plastic-ban

15 Nationally Appropriate Mitigation Strategies (NAMA). "Costa Rica's climate change strategy." Available at: www.namacafe.org/en/costa-ricas-climate-change-strategy

16 Constitution Project. Costa Rica's constitution. Available at: www.constituteproject. org/constitution/Costa_Rica_2011?lang=en

17 IFLscience. "Costa Rica has only used renewable energy for electricity this year." Available at: www.iflscience.com/environment/costa-rica-has-only-used-renewables-electricity-year/

18 "List of power stations in Costa Rica." *Wikipedia.* Available at: https://en.wikipedia. org/wiki/List_of_power_stations_in_Costa_Rica

19 *Nationally Appropriate Mitigation Actions.* "Costa Rica's climate change strategy." Available at: www.namacafe.org/en/costa-ricas-climate-change-strategy

20 Cory Doctorow. "The Vulnerable 20 Group Coalition." Available at: https://boing boing.net/2016/05/31/the-vulnerable-20-group-coali.html

21 The World Bank. "Vulnerable twenty ministers call for more action and investment in climate resiliency and low-emissions development." Available at: www.world bank.org/en/news/feature/2015/10/08/vulnerable-twenty-ministers-more-action-investment-climate-resilience-low-emissions-development

22 Track0.org. "48 Climate Vulnerable Forum count to 100% renewables at COP22." Available at: http://track0.org/2016/11/48-climate-vulnerable-forum-countries-commit-to-100-renewables-at-cop22/. Climate Home. "Vulnerable nations call on G-20 to end fossil fuel subsidies by 2020." Available at: www.climatechangenews. com/2017/04/24/vulnerable-nations-call-g20-end-fossil-fuel-subsidies-2020/

23 Alliance of Small Island States. "About." Available at: http://aosis.org/about/ members/

24 John W. Ashe, Robert Van Lierop, and Anila Cherian. "The role of the Alliance of Small Island States (AOSIS) in the negotiation of the United Nations Framework Convention on Climate Change (UNFCCC)." *Natural Resources Forum*, 23(3), August 1999. Available at: http://onlinelibrary.wiley.com/doi/10.1111/j.1477-8947.1999.tb00910.x/abstract

25 *New York Times.* "The Marshall Islands are disappearing." Available at: www. nytimes.com/interactive/2015/12/02/world/The-Marshall-Islands-Are-Disappear ing.html?mcubz=0&_r=0

26 US Department of State. "U.S. relations with Marshall Islands." Available at: www. state.gov/r/pa/ei/bgn/26551.htm

27 Climate Analytics. "Half a degree would make a world of difference." Available at: http://climateanalytics.org/blog/2017/half-a-degree-could-make-a-world-of-difference.html

28 Climate Central. "These cities are most vulnerable to major coastal flooding and sea rise," October 25, 2017. Available at: www.climatecentral.org/news/us-cities-most-vulnerable-major-coastal-flooding-sea-level-rise-21748

CHAPTER 9

Proposals and Solutions

It can be said that we are headed toward doomsday—a disastrous and deadly fate as the Bulletin of Atomic Scientists reminded the world's people on January 25, 2018. They cited climate change as a major reason. The earth is heating up at a pace that makes it difficult to control or stop. People are told that rising temperatures will destroy plant and animal habitats, reduce yields of important food crops, and create the conditions for fierce storms. People will be exposed to the ravages of flooding and drought, and some parts of the globe will be uninhabitable. It seems so inevitable.

Yet the entire world, with the exception of the United States, is moving along at a fast clip to end fossil fuels and rely solely on renewables. (To put it dramatically, but correctly, Trump could end the world as we know it.) There are extraordinary innovations that aim to reverse or slow heating. Yes, this is very much like the early days of the Industrial Revolution or the development of the printing press. The floodgates are open to new ideas and exciting inventions. The big difference between the Industrial Revolution and what we can call the "renewable revolution" is that competition was the rule in the Industrial Revolution, whereas everyone, every country, and every community needs to—must—cooperate to save the planet. The spirit has been one that promotes openness and collaboration as well as public involvement. This chapter focuses on big projects—indeed, very big projects—and the next chapter on projects that all of us can do. Indeed, they make a difference if each of us does them, and the doing of them unites the world's peoples.

STRAIGHTFORWARD SOLUTIONS AND SOME COMPLICATED ONES

Tide Power

The gravitational pull of the moon and sun along with the rotation of the earth cause the tides. In some places, tides cause water levels near the shore to vary up to 40 feet. People harnessed this movement of water to operate grain mills more than a thousand years ago in Europe. Today, tidal energy systems generate electricity. Producing tidal energy economically requires a tidal range of at least 10 feet.[1] Tidal turbines, which look similar to wind turbines, can be placed on the seabed where there is strong tidal flow. Because water is about 800 times denser than air, tidal turbines have to be much sturdier and heavier than wind turbines. Turbines have been built in Canada, France, Scotland, and South Korea. Others are under construction.[2]

Wave Power

In many parts of the world, the wind blows with enough consistency and force to provide continuous waves along the shoreline. Ocean waves contain tremendous energy potential, and a machine able to exploit wave power is known as a wave energy converter (WEC), which can be installed close to shore or in deeper water.[3] There are several types. They differ in their engineering and design, and they differ in their orientation to the waves or in the manner in which they convert energy from the waves. WECs are more efficient than wind turbines.[4]

Garbage Power (Waste-to-Energy)

This is the conversion of non-recyclable waste (garbage) into usable heat, electricity, or fuel. There are nearly 90 facilities in the United States that recover energy from the combustion of municipal solid waste. They burned about 29 million tons of waste in 2015, generated nearly 14 billion kilowatt hours of electricity, having the capacity to produce electricity for 2,720 megawatts of power. (To give a better idea, the Pinella, Florida, facility processes up to one million tons of waste every year and produces up to 75 megawatts of electricity every hour to 45,000 homes and businesses.)[5] The process is simple enough: burning waste heats water into steam that drives a turbine to create electricity. The process can reduce a community's landfill volume by up to 90%, and prevent one ton of carbon dioxide release for every ton of waste burned.[6]

Bicycle Super-Highways

These connect towns and cities for cyclists to commute to work or simply to ride for pleasure. Denmark, Germany, and Sweden have each created one.[7] The one in Greater Copenhagen opened in 2012 and has amenities such as air pumps, safe intersections, and traffic lights timed to average cycling speed, reducing the number of stops.[8] Yes, keeping automobiles off the roads goes a long way to slowing global warming.

Seasteading

As the sea rises, one solution has been to fortify homes, land, and the coast by building walls against the rising water. But in the long run this does not work since the sea keeps on rising. Seasteading—or building homes and towns on the water—is the superior solution. Seasteading is popular in many countries: Netherlands,[9] Denmark,[10] Sweden,[11] Norway,[12] China,[13] New Zealand,[14] Greece,[15] Qatar,[16] India,[17] Germany,[18] and among the Uros (an indigenous tribe in Bolivia).[19] But it has not yet caught on in the U.S., at least to the extent that it has elsewhere.[20]

Hydroponics

This is the method of growing plants without soil and in nutrient-rich water solvent, and because it usually involves arranging plants vertically, hydroponic farming can be carried out in cities—in, say, warehouses or garages. Plants may be grown with only their roots exposed to the mineral solution, or the roots may be supported by an inert medium, such as gravel. The nutrients in hydroponics can come from different sources, including fish waste, duck manure, or mineral nutrients.[21]

Bamboo Houses

Bamboo is a highly sustainable building material: it is grown without pesticides or chemical fertilizers; requires no irrigation; rarely needs replanting; grows rapidly and can be harvested in 3–5 years; produces 35% more oxygen than equivalent stand of trees; and it sequesters carbon dioxide and is carbon neutral.[22]

ELIMINATING CO_2

Our goal—the world's goal, the main goal—is zero CO_2 emissions by 2050. That means unless emissions are entirely balanced by, say, trees (which is unlikely), we (the world's peoples) must eliminate CO_2. If we don't, there is a risk that the earth will warm as much as 4.5°C by the end of this century, with the chance that this will reach 6°C overall and on average. Of course, these estimates do not account for the great variation around the globe—that is to say, the residents of Finland may experience summers that are hot, but tolerable, while the Persian Gulf is likely to be uninhabitable by 2100.[23] We can only speculate about the implications of this for migration, refugees, conflict, and hunger.

Carbon Tax

The Paris Agreement, according to some economists, will fail because there are no binding cooperative agreements. For this reason, a tax is advocated that directly sets a price on greenhouse gas emissions or on the carbon content of fossil fuels.[24] This helps to shift the burden for the damage back to those who are responsible for it and to those who can reduce it. Instead of dictating who should reduce emissions (and where and how), a carbon price gives an economic signal and polluters decide for themselves whether to discontinue their polluting activity, reduce emissions, or continue polluting and pay for it. In this way, the overall environmental goal is achieved in the most flexible and least-cost way to society. The carbon price also stimulates clean technology and market innovation, fuelling new, low-carbon drivers of economic growth.[25]

Ocean Carbon Sinks

One possibility to eliminate carbon dioxide from the atmosphere is to fertilize the oceans with iron or other nutrients to enable plankton to become more efficient carbon sinks. Studies have established the important role of plankton networks in removing carbon from the atmosphere and depositing it deep in the ocean. And it opens up opportunities for caring for the ocean in ways that encourage the ocean to absorb more carbon.[26] This is important because oceans are losing their capacity to absorb carbon dioxide.[27]

Carbon Capture and Storage (Sequestration) or CCS

There are two primary types of carbon sequestration or Carbon Capture and Storage (CCS). One type involves capturing carbon dioxide at its source (e.g., power plants, industrial processes) and subsequently putting it into deep, long-term storage in non-atmospheric reservoirs (e.g., depleted oil and gas reservoirs, coal seams, deep saline formations, or the ocean).[28] The other type of carbon sequestration is terrestrial sequestration and it involves enhancing natural processes such as forestation to increase the removal of carbon from the atmosphere. Besides enhancing forests, it includes land management practices that maximize the amount of carbon that remains stored in the soil and plant material for the long term. No-till farming, wetland management, rangeland management, and reforestation are examples of terrestrial sequestration practices that are already in use.[29]

Cap and Trade, or Emissions Trading Systems (ETS)

These cap the total level of greenhouse gas emissions, aimed most at reducing CO_2 and allows those industries with low emissions to sell their extra allowances to larger emitters. By creating supply and demand for emissions allowances, an ETS establishes a market price for greenhouse gas emissions. The cap helps ensure that the required emission reductions will take place to keep the emitters within their pre-allocated carbon budget. The first mandatory Cap and Trade mandate in the U.S. is the Regional Greenhouse Gas Initiative (RGGI). It is a cooperative effort among the states of Connecticut, Delaware, Maine, Maryland, Massachusetts, New Hampshire, New York, Rhode Island, and Vermont to cap and reduce CO_2 emissions.[30]

PERFORMANCE-BASED NAVIGATION (PBN)

PBN is navigation for airplanes that uses global navigation satellite systems (GNSS) and computerized on-board systems. Because it allows optimum and flexible routing, it reduces CO^2 emissions.[31] This is in contrast with traditional sensor-specific navigation based largely on fixed ground-based beacons guiding aircraft along published routes via waypoints defined by these beacons.[32] Bangladesh was a pioneer in advancing this,[33] and it has become increasingly clear in the international aviation community that reducing emissions is imperative.[34]

Direct Air Capture (DAC)

DAC is an emerging class of technologies capable of separating carbon dioxide (CO_2) directly from the air. DAC systems can be thought of as artificial trees. Whereas trees extract CO_2 from the air using photosynthesis,

DAC systems extract CO_2 from the air using chemicals that bind to CO_2 but not to other atmospheric chemicals (such as nitrogen and oxygen). As air passes over the chemicals used in DAC systems, CO_2 "sticks" to these chemicals. When energy is added to the system, the purified CO_2 "unsticks" from the chemicals, and then the chemicals can be redeployed to capture more CO_2 from the air. The idea of separating CO_2 from air is not new, and has been done on submarines and in space applications for decades. Yet large-scale DAC systems are just being developed. Today, there are four leading commercial DAC system development efforts, along with one academic center pursuing DAC research.[35]

Converting Carbon Dioxide

At least two companies—NRG and Adidas—have made plastic shoes out of CO_2.[36] Adidas's shoes cost $200 a pair. A pair will use, on average, 11 plastic bottles and include recycled materials for the laces, heel lining, and sock-liner covers.[37] A factory in Tuticorin, India, is capturing CO_2 from its own coal-powered boiler and using it to make baking soda.[38]

ANOTHER CULPRIT BESIDES CO$_2$: NITROUS OXIDE

Nitrous oxide emissions are primarily produced when nitrogen is added to the soil through the use of fertilizers, and are emitted during the breakdown of nitrogen in livestock manure and urine. Two other major sources include fuel combustion and industrial production. Nitrous oxide has a global warming potential of 310 times that of CO_2 over a hundred-year timescale. In addition to contributing to climate change, nitrous oxide contributes to ozone depletion. In international climate talks in 2017, attention focused on reducing nitrous oxide emissions in two ways: One is through dietary changes: eating less meat would help to reduce emissions, given that the livestock industry accounts for a large share of the nitrous oxide emissions. The second is a better management of livestock, croplands, and integrated agro-forestry systems.[39]

According to the World Bank, additional environmental and health benefits of cutting nitrous oxide emissions, including improved water quality and air quality, would amount to US$160 billion per year. In addition to improving nitrogen use efficiency, the World Bank suggests the following policies to unlock this potential: targeting fertilizer subsidies to avoid the overuse of nitrogen fertilizers; investment in soil conservation; improved cook stoves; and improvements in waste management.[40]

ANOTHER CULPRIT BESIDES CO$_2$: HYDROFLUOROCARBONS (HFCS)

Hydrofluorocarbons (HFCs) are by far the most common of the Fluorinated Gases (F-gases). [41] HFCs are man-made greenhouse gases that are used primarily in refrigeration and air-conditioning units for homes, buildings and industrial operations, which accounts for about 79% of total HFC use, and aerosols account for an additional 5%.

Many HFCs are highly potent warming agents, some with global warming potential of over 12,000 times that of CO$_2$. The production, consumption, and demand for refrigerators and air-conditioners is growing fast worldwide, primarily in developing countries. According to the Climate and Clean Air Coalition (CCAC), the HFC production, consumption and emissions are growing at a rate of 8% per year, and up to 15% in some countries. The Climate and Clean Air Coalition said that HFCs could contribute up to 0.1°C warming by 2050 and up to 0.5°C warming by 2100. It was highlighted at a 2016 UNFCCC meeting that possible future HFC emissions are a significant obstacle to the 450 ppm stabilization target, and that HFCs could account for as much as 20% of all greenhouse gas emissions by 2030 if present trends continue.[42]

Apparently, the announcement at the October 2016 Kigali climate meetings mobilized action and nearly 200 countries struck an important agreement that, by the late 2040s, all countries will significantly reduce hydrofluorocarbons.[43] And member states of the EU are committed to cut fluorinated gases by two-thirds by 2030.[44] This is encouraging news since it is estimated that adequate action on HFCs could prevent up to 2 billion tons of CO$_2$ equivalent emissions over the next decade, and over 100 billion tons of CO$_2$ emissions by 2050.[45]

ANOTHER CULPRIT BESIDES CO$_2$: METHANE

Released by cows as a result of their digestive processes, methane accounts for 25% of all emissions, and methane happens to be one of the worst greenhouse gases. A recent study reports that beef compared with poultry, pork and eggs per gram of protein, use 28 times more land, 11 times more irrigation water for feed, and release five times more greenhouse gases linked to global warming.[46] Besides, cows are responsible for more biogas— namely the byproduct of manure. One method to capture the gases of raw manure is to heat the manure in an aerobic digester, which is a heated, oxygen- free container that essentially continues the digestion that began in the cow's stomach. Capturing biogas from cattle, hog and poultry farms can reduce greenhouse gas emissions and recovering the methane from the biogas can provide a cost-effective source of renewable energy. Cow dung without its gas can be used for fertilizer or potting soil.[47]

Seaweed

Seaweed diet for cows reduces methane by 99%.[48] A Canadian farmer, Joe Dorgan, inadvertently conducted an experiment on his herd in Prince Edward Island, Canada. Dorgan noticed that his cows that grazed on washed-up seaweed in paddocks along the shore were healthier and more productive than those that stayed in the field. He began feeding his cows a mixture of local storm-tossed seaweed and found the new diet saved him money. Dorgan sold his cows and went into the business of making and selling feed made up partly of seaweed.[49]

Harvesting methane

Another approach to reduce methane is to put plastic backpacks on cows. The backpack is attached to tubes inserted into the cow's rumens (their biggest digestive tract), and then it captures and collects the gases emitted through the cow's mouth or intestinal tract via a tube inserted through the cow's skin (which the researchers claim is painless). The gas is then condensed and ready to use to provide power for activities such as cooking, lighting a home, or even driving a car. The methane that is extracted over the course of a day—about 300 liters—is enough to run a car for 24 hours.

PLASTIC: A HUGE CRISIS

A million plastic bottles are bought around the world every minute and the number will jump another 20% by 2021, creating an environmental crisis. Between 5 and 13 million tons of plastic leak into the world's oceans each year to be ingested by seabirds, fish and other organisms, and by 2050 the ocean will contain more plastic by weight than fish.[50] Since plastic is inert, it does not contribute to global warming. However, its production does contribute to global warming. According to the Pacific Institute, the production of recyclable plastic bottles that water comes in requires about 17 million barrels of oil to make. That's enough oil to fuel more than 1 million U.S. cars per year. It also takes water to bottle water, requiring manufacturers to use about 3 liters of water for every liter of bottled water produced.[51]

Great Pacific Garbage Patch

This immense area in the North Pacific is made up mostly of plastic and micro-plastic pieces (which are a combination of "micro-beads" found in cosmetics. It is estimated that it is 700,000 square kilometers or 270,000 square miles (about the size of Texas) with the periphery spanning further for a total of 15,000,000 square kilometers (5,800,000 square miles).[52] The dimensions of this morass of waste are continually morphing, caught in one of the ocean's huge rotating currents. These currents have trapped and accumulated a soup of plastic waste, including large items and smaller broken-down micro plastics that can be eaten by fish and enter the food chain.

Ocean Cleanup

A Dutch foundation devoted to rid the oceans of plastic, has announced it will start extracting plastic from the Great Pacific Garbage Patch within the next 12 months. Ocean Cleanup further announced that parts of its first cleanup system are already in production. The main idea behind the Ocean Cleanup is to let the ocean currents do the work. An installation of U-shaped screens channel floating plastic to a central point. The concentrated plastic can then be extracted and shipped to shore for recycling into durable products. Ocean Cleanup proposes the introduction of a mobile, or drifting, system of floating screens affixed to anchors.[53] Ocean Cleanup's models indicate that a full-scale system roll-out could clean up 50% of the Great Pacific Garbage Patch in five years.

Biodegradable Plastic

Biodegradable plastics are made from all-natural plant materials. These can include corn oil, orange peels, starch, and plants. Traditional plastic is made with chemical fillers that can be harmful to the environment when released when the plastic is melted down. But from biodegradable plastic comes a substance made from natural sources that does not contain these chemical fillers, and does not pose the same risk to the environment.[54] This is controversial, however, and some studies do not support the conclusion.[55]

"Eating Plastic"

Scientists recently found that the ordinary wax worm, given the opportunity, will eat plastic and as it does so these are transformed into ethylene glycol, which is commonly used in antifreeze. These are very preliminary discoveries, yet plausible because wax and polyethylene are chemically similar. If this pans out, it would be very exciting. "We could put wax worms to work while they dine at their leisure."[56]

Plastic Roads

These have been constructed from plastic bags, disposable cups, and plastic bottles, all of which are collected from garbage dumps. When mixed with hot bitumen, plastics melt to form an oily substance and the mixture is laid on the road surface like a normal tar road. Comparatively, they have the following advantages: they involve faster construction and less maintenance time; they are found to be of higher quality with a longer lifespan and require little to no maintenance; and are virtually impervious to conditions such as the weather and weeds.[57] They are currently being developed in the U.S.,[58] Britain,[59] India, [60] and the Netherlands.[61]

Plastic Houses

Conventional building materials have drawbacks. Wooden structures deplete the stock of trees. The production of concrete generates a great amount of carbon dioxide and the same is the case with making bricks. One creative solution is to build houses from recycled plastic. This is a big plus because it is cheap and because recycling plastic keeps it from accumulating in the ocean. A Colombian architect, Oscar Mendez, with his company, Conceptos Plásticos, launched this concept a couple of years ago and the homes they build cost less than $6,000, providing very affordable dwellings for poor and working-class Colombians.[62]

GEO-ENGINEERING

Geo-engineering is large-scale manipulation of the environment, and, therefore, extremely controversial. There are many examples, some completed, but most are proposed.

Solar Power Plant

Crescent Dunes in Tonopah, Nevada, aims to simultaneously produce the cheapest solar thermal power and to dispatch that power for up to ten hours after the sun sets. It uses over 10,000 mirrors to focus sunlight on a heat receiver atop a 165-meter high tower. It heats a molten mixture of nitrate salts that can be stored in insulated tanks and withdrawn on demand to run the plant's steam generators and turbine when electricity is most valuable.[63]

Synlight

Synlight is located in Juelich, Germany, about 30 kilometers (19 miles) west of Cologne.[64] Designed to produce hydrogen, it is a giant honeycomb-like set-up of 149 spotlights that will produce light that is about 10,000 times the intensity of direct sunlight. Synlight currently uses a vast amount of energy, and when all the lamps are swiveled to concentrate light on a single spot, the instrument can generate temperatures of around 3,500°C—around two to three times the temperature of a blast furnace. If successful, it will produce carbon-neutral hydrogen that can be used as a fuel for airplanes and cars. Many consider hydrogen to be the fuel of the future because it produces no carbon emissions when burned, and therefore doesn't add to global warming. But while hydrogen is the most common element in the universe, it is rare on earth.[65]

Limestone Particles

Distributing limestone particles into the upper atmosphere could slow global warming because it would reflect and scatter incoming solar radiation, while repairing the ozone layer.[66] It is a theoretical and controversial approach to reducing some of the impacts of climate change by reflecting a small amount of inbound sunlight back out into space. It is in the early stages of research, but it is already a controversial topic.

Solar Radiation Management

Different SRM techniques have been proposed, but the proposals receiving the most attention from researchers would involve brightening marine clouds by spraying seawater into the lower atmosphere, or replicating the cooling effect of volcanoes by spraying reflective sulfate particles into the upper atmosphere (the stratosphere).[67]

Block the Sun with Mirrors

This was proposed by Lowell Wood of Lawrence Livermore National Laboratory in the early 2000s, though he cautioned that the mirror should be considered only as a measure of last resort. Why? Because the mirror would have to have an area of 600,000 square miles—a slightly smaller area than Greenland—and launching something that big would be extremely expensive. Another option is to have billions of smaller mirrors, which was proposed by Roger Angel of the University of Arizona in 2006. In either case, the mirror or mirrors would block just 1–2% of the sun's light, but that would be enough, it is said by advocates of either scheme, to cool the planet.[68]

Freeze the Arctic

Desperate times call for desperate measures, and with temperatures near the North Pole reaching 2°C (36°F), ice volume was at record low in 2017.[69] The loss of the Arctic's summer sea ice cover will disrupt life in the region, endanger many of its species, from Arctic cod to polar bears, and destroy a pristine habitat. It will also trigger further warming of the planet because the loss of sea ice cover, which reflects solar radiation back into space, will disrupt weather patterns across the northern hemisphere and melt permafrost, releasing more carbon gases into the atmosphere. With less ice to reflect solar radiation back into space, the dark ocean waters of the high latitudes will warm and the Arctic will heat up even further. What will have the most significant impact on all humans will be sea rise.[70] One prediction is that sea rise will be 216 feet, affecting the entire world.[71] A team of scientists at Arizona State University has come up with a novel solution. They propose building 10 million wind-powered pumps over the Arctic ice cap. In winter, these would be used to pump water to the surface of the ice where it would freeze, thickening the ice cap, adding about an extra 3 feet to the existing ice.[72]

Manipulate the Clouds

Cirrus clouds are thin and wispy that form at high altitudes and do not reflect much solar radiation back into space, creating a greenhouse effect. The higher the altitude at which they form, the larger the warming effect on the climate. And in a warmer climate, cirrus clouds form at higher altitudes, up to 20 kilometers above the earth's surface, and because they let much sunlight pass through, they may trap the earth's heat, just as greenhouse gases do. Therefore, they have a net warming effect that helps magnify warming. These clouds could be thinned out, leading to a reduction in their warming effect, by seeding them with aerosol particles like sulfuric or nitric acid. Therefore, if these clouds were injected with such particles it would result in clouds that have less of a warming effect. So far, the risks are considered too great to try.[73]

SOLAR

We are all familiar with the huge success of solar. For example, half the power needs for the state of California is met by solar, and as the price has dramatically declined, the number of installations in the U.S. and European countries has soared. There are other uses of photoelectric installations.

Solar Roads

The U.S.[74] and France[75] have experimented with solar roads, and the Netherlands with solar bike paths.[76] Idaho-based Solar Roadways has received three rounds of U.S. government funding (plus $2 million in venture capital) to test its technology.[77]

Solar Tents

In a remote highland community in Central Bolivia the conditions for growing food are extremely harsh, if not forbidding. Indigenous women were provided with solar tents in which to grow vegetables. As a result, each of their family members has protein, minerals, and vitamins.[78]

OTHER INNOVATIONS

Solid Rain

Solid rain is a potassium-based powder which is capable of absorbing water up to 500 times its size. Solid rain acts as a personal underground reservoir that retains water in the roots of any plant. This retained water is then slowly dispersed in the soil, keeping the plant constantly hydrated.[79] It was invented by Sergio Jésus Rico Velasco, a Mexican chemical engineer who spent decades trying to mitigate his country's drought issues.[80]

Kinetic Energy

Kids' feet light up the soccer field in Brazil as they pound the grass field[81] and as they run. In other words, the kinetic energy triggers the solar panels. So successful, Nigeria has adopted the same technology for a soccer field in Lagos.

Living Buildings

The concept of the "living building" is advocated as a new ideal for design and construction. It is a structure that generates all of its own energy with renewable nontoxic resources, captures and treats all of its water, and operates efficiently and for maximum beauty, with spaces that connect occupants to light, air, food, nature, and community.[82] The concept was developed by the Living Future Institute with the aim that buildings are self-sufficient and they remain within the resource limits of their site. Living Buildings produce more energy than they use and collect.[83]

Spray-on Silicon

A promising new film technology is named "Perovski cells" after a 19th-century Russian mineralogist, Lev Perovski. Unlike silicon-based photovoltaic (PV) cells used in solar panels, Perovski cells are soluble in a variety of solvents, so can be easily sprayed on to a surface, similar to inks or paints. They have promise as replacing solar cell technology.[84]

Hydrogen-powered Catamaran

A boat named *Energy Observer*[85] has started on its journey to circumnavigate the world using only renewable energy sources and desalinated ocean water. Not only is hydrogen created to help power the boat, but it also has on board a giant kite and solar panels.

CONCLUSIONS

Extraordinary efforts are being undertaken to avert a global crisis. Some efforts, of course, are risky, but others are not. It is extremely important, whenever possible, to engage young people in these efforts since they will be most affected by the likely calamities due to warming of the earth. What is truly astonishing is the incredible innovation taking place. Proposals include blocking the sun, manipulating the clouds, and freezing the Arctic. And yes, there are some challenges that are not so esoteric but absolutely necessary. We must collectively abandon fossil fuels and adopt renewables.

NOTES

1   *National Geographic.* "Tidal energy." Available at: www.nationalgeographic.org/encyclopedia/tidal-energy/

2   Tidal turbines are more expensive to build than wind turbines, but capture more energy with the same size blades. US Energy Information Administration. "Energy explained." Available at: www.eia.gov/energyexplained/index.cfm?page=hydro power_tidal; Wikipedia. "List of Tidal Power Stations": https://en.wikipedia.org/wiki/List_of_tidal_power_stations

3   AE News. "Wave power." Available at: www.alternative-energy-news.info/tech nology/hydro/wave-power/

4   Drawdown. "Coming attractions: Solid state wave energy." Available at: www. drawdown.org/solutions/coming-attractions/solid-state-wave-energy

5   "Waste to energy incineration." Available at: http://ecocloud.sustainablesv.org/index.php/topics/materials/diverting-waste/waste-to-energy/incineration

6   Deltaway. "Waste to energy: How it works." Available at: www.deltawayenergy. com/wte-tools/wte-anatomy/

7   Phys.org. "Germany gives green light to bicycle highways." Available at: https:// phys.org/news/2015-12-germany-green-bicycle-highways.html.   Treehugger. "A new four-lane superhighway to be built only for bikes." Available at: www. treehugger.com/bikes/new-cycling-superhighway-not-us.html

8   Denmark. "Cycle superhighway." Available at: http://denmark.dk/en/green-living/bicycle-culture/cycle-super-highway

9   DW. "Floating houses to fight climate change in Holland." Available at: www. dw.com/en/floating-houses-to-fight-climate-change-in-holland/a-17532376

10   Seasteading Institute. Available at: www.seasteading.org/2012/09/ambassador-lasse-birk-olesen-at-tedx-copenhagen-seasteading-technology-politics/

11   "Contemporary floating home in Sweden." Available at: www.homedsgn.com/2012/03/24/contemporary-floating-home-in-sweden/

12   "Luxury floating house 200 sqm with seaview." Available at: www.airbnb.com/rooms/13458479

13   Linkedin. "Ark Hotel China is floating hotel unique in the world." Available at: www.linkedin.com/pulse/ark-hotel-china-floating-unique-world-shekhar-gupta

14   Stuff. "Floating homes come to Auckland." Available at: www.stuff.co.nz/life-style/home-property/73793121/floating-homes-coming-to-auckland

15   George Kottas. "Floating house in Vegoritis Lake Pella Greece." Available at: https://500px.com/photo/85544901/floating-house-in-vegoritis-lake-pella-greece-by-george-kottas

16   Inhabitat. "Qatar unveils luxurious off-grid floating hotels for 2022 world cup." Available at: http://inhabitat.com/qatar-unveils-luxurious-off-grid-floating-hotels-for-2022-world-cup/

17   MGS Architecture. "Floating and moving houses." Available at: www.mgsarchi tecture.in/projects/380-floating-and-moving-houses-a-need-of-tomorrow.html

18   Dualdocker. "Floating houses in Germany." Available at: www.dualdocker.com/videos-links/floating-houses-in-germany.html

19   Messy Nessy. "The floating straw islands of Uros." Available at: www.messy nessychic.com/2015/05/13/the-floating-straw-islands-of-uros-an-ancient-tribe-surviving-in-a-tourist-hub/

20   Available at: www.seasteading.org/

21   Paul Marks. "Vertical farms sprouting all over the world." *New Scientist*, January 15, 2014. Available at: www.newscientist.com/article/mg22129524-100-vertical-farms-sprouting-all-over-the-world/

22 Econation. "Bamboo." Available at: https://econation.co.nz/bamboo/; "Bamboo living." Available at: http://bambooliving.com/index.php/52-bamboodist-ezine

23 Deep Decarbonization Pathways Project (DDPP). Available at: http://deepdecarbon ization.org/about/

24 Carbon Tax Center. "What's a carbon tax?" Available at: www.carbontax.org/ whats-a-carbon-tax/

25 Carbon-Price.com. "Global carbon pricing." Available at: http://carbon-price.com/

26 Wynne Parry. "Could fertilizing the oceans reduce climate change?" *Live Science*, July 18, 2012. Available at: www.livescience.com/21684-geoengineering-iron-fertilization-climate.html

27 Scripps Institution of Oceanography. "The Keeling Curve: How much CO_2 can the oceans take up?" July 3, 2013. Available at: https://scripps.ucsd.edu/ programs/keelingcurve/2013/07/03/how-much-co2-can-the-oceans-take-up/

28 There are three stages to CCS: capture, transport, and safe underground storage. 1. Capture—First, the carbon dioxide is removed, or separated, say, from coal and gas power plants. 2. Transport—The carbon dioxide is then compressed and transported to a suitable storage site. 3. Storage—The carbon dioxide is then injected into a suitable storage site deep below the ground. The storage site must be a geological formation that ensures safe and permanent storage. Carbon Capture and Storage Association. "Frequently asked questions." Available at: www.ccsassocia tion.org/faqs/ccs-general/

29 Plains CO_2 Reduction (PCOR) Partnership. "What is CO_2 sequestration?" Available at: www.undeerc.org/pcor/Sequestration/WhatIsSequestration.aspx. Perennials' longer roots allow them to hold onto moisture and carbon, making soil richer, even during droughts and floods of the past two years in states like Kansas and California that many scientists have linked to climate change.

30 Regional Greenhouse Gas Initiative. Available at: www.rggi.org/

31 Djordje Samardzija, "Climate based navigation – A new way to reduce aviation CO_2 emissions." Available at: www.halasesar.net/sites/default/files/documents/ p210-samardzija.pdf

32 Australian Government. Civil Aviation. "What is performance based navigation (PBN)?" Available at: www.casa.gov.au/standard-page/what-performance-based-navigation-pbn

33 Inter-press Service. "Flying green in Bangladesh." Available at: www.ipsnews.net/ 2017/05/flying-green-in-bangladesh/

34 International Civil Aviation Organization. Assembly – 39th Session. Executive Committee. "Agenda Items 20 and 22. Environmental Protection. Available at: www.icao.int/Meetings/a39/Documents/WP/wp_170_en.pdf

35 Center for Carbon Renewal. "Direct air capture explained in 10 questions." Available at: www.centerforcarbonremoval.org/blog-posts/2015/9/20/direct-air-capture-explained-in-10-questions

36 Cool Hunting. "NRG's sneaker composed of repurposed CO_2." Available at: www.coolhunting.com/tech/nrg-co2-sneaker; *Anthropocene*. "How to turn CO_2 into fuel, drugs, and running shoes." Available at: www.anthropocenemagazine.org/ 2015/09/what-if-we-solved-our-co2-problem-by-making-stuff-with-it/

37 Min Liu et al., "Enhanced electrocatalytic CO_2 reduction via field-induced reagent concentration." *Nature*, 537: 382–386, September 15, 2016. Available at: www. nature.com/nature/journal/v537/n7620/full/nature19060.html; see: Marit Mitchel,

"Recycling carbon dioxide: U of T researchers efficiently reduce climate-warming CO_2 into building blocks for fuels." *Engineering News*. Available at: http://news.engineering.utoronto.ca/recycling-carbon-dioxide-u-t-researchers-efficiently-reduce-climate-warming-co2-building-blocks-fuels/

38   Roger Harriban, "Indian firm makes carbon capture breakthrough," January 3, 2017. Available at: www.theguardian.com/environment/2017/jan/03/indian-firm-carbon-capture-breakthrough-carbonclean

39   United Nations Framework Convention on Climate Change. "Experts say Paris can help cut non-CO_2 gases." Available at: http://newsroom.unfccc.int/unfccc-newsroom/bonn-meeting-identifies-ways-to-curb-non-c02-gases/

40   The World Bank. "Minding the stock: Bringing public policy to bear on livestock sector development." 2009. Available at: http://siteresources.worldbank.org/INTARD/Resources/FinalMindingtheStock.pdf

41   The other fluorinated gases are: 1) Perfluorocarbons (PFCs), which are compounds consisting of fluorine and carbon and used in the electronics, cosmetics, and pharmaceutical industries, as well as in refrigeration when combined with other gases. PFCs were commonly used as fire extinguishants in the past and are still found in older fire protection systems. They are also a by-product of the aluminum smelting process. 2) Sulphur hexafluoride (SF6) is used primarily as an insulation gas.

42   United Nations Framework Convention on Climate Change. UN Climate Change Newsroom. "Experts say Paris can help cut CO_2 gases." Available at: http://newsroom.unfccc.int/unfccc-newsroom/bonn-meeting-identifies-ways-to-curb-non-c02-gases/

43   *The Guardian*, October 15, 2016. "Kigali deal on HFCs is a big step in fighting climate change." Available at: www.theguardian.com/environment/2016/oct/15/kigali-deal-hfcs-climate-change; *Scientific American*, October 17, 2016. "World leaders agree to phase out heat-trapping hydrofluorocarbons." Available at: www.scientificamerican.com/article/world-leaders-agree-to-phase-out-heat-trapping-hydrofluorocarbons/

44   European Commission. "Fluorinated greenhouse gases." Available at: https://ec.europa.eu/clima/policies/f-gas_en

45   United Nations Framework Convention on Climate Change. UN Climate Change Newsroom. "Experts say Paris can help cut CO_2 gases." Available at: http://newsroom.unfccc.int/unfccc-newsroom/bonn-meeting-identifies-ways-to-curb-non-c02-gases/

46   United Nations Framework Convention on Climate Change. UN Climate Change Newsroom. "Go easy on the beef for the climate." Available at: http://newsroom.unfccc.int/nature-s-role/go-easy-on-the-beef-for-the-climate/

47   Environmental Protection Agency. "Learn about biogas recovery." Available at: www.epa.gov/agstar/learn-about-biogas-recovery. America's Dairyland. "How do digesters work?" Available at: www.americasdairyland.com/on-the-farm/environment/bioenergy/howdigesterswork. Thinkprogress. "How farms across America are using cow manure for renewable energy." Available at: https://thinkprogress.org/how-farms-across-america-are-using-cow-manure-for-renewable-energy-b49e9ff06e92. Phys.org. "Turning cow poo into power is profitable." Available at: https://phys.org/news/2015-08-cow-poo-power-profitable-farm.html

48 *Toronto Sun*, February 12, 2017. "How a seaweed diet dramatically cuts cows methane output." Available at: www.torontosun.com/2016/12/02/how-a-sea weed-diet-dramatically-cuts-cows-methane-output. Foodtank News. "Cows fed seaweed contribute less methane emissions to atmosphere." Available at: https://food tank.com/news/2017/06/seaweed-reduce-cow-methane-emission/. Fast Company. "These backpacks collect their fart gas and store it for energy." Available at: www. fastcompany.com/3028933/these-backpacks-for-cows-collect-their-fart-gas-and-store-it-for-energy

49 AllAboutFeed. "Seaweed makes healthier cows." Available at: www.allaboutfeed. net/New-Proteins/Articles/2016/12/Seaweed-makes-healthier-cows-69832E/

50 Sandra Laville and Matthew Taylor. "A million bottles a minute: World's plastic binge 'as dangerous as climate change.'" *The Guardian*, June 28, 2017. Available at: www.theguardian.com/environment/2017/jun/28/a-million-a-minute-worlds-plastic-bottle-binge-as-dangerous-as-climate-change

51 Pacific Institute. "Bottled water and energy fact sheet." Available at: http://pacinst. org/publication/bottled-water-and-energy-a-fact-sheet/

52 *National Geographic*. Encylopedic Entry. "Great Pacific Garbage Patch." Available at: www.nationalgeographic.org/encyclopedia/great-pacific-garbage-patch/

53 The Ocean Cleanup. Available at: www.theoceancleanup.com/press/press-releases-show/item/the-ocean-cleanup-announces-pacific-cleanup-to-start-in-2018/

54 Precision Engineered Products. Biodegradable Plastics. Available at: www.pepct plastics.com/resources/connecticut-plastics-learning-center/biodegradable-plastics/

55 UN Environment Program. "Biodegradable plastics are not the answer to reducing marine litter." Available at: www.unep.org/asiapacific/biodegradable-plastics-are-not-answer-reducing-marine-litter-says-un

56 *Scientific American*. "Plastic eating worms could inspire waste degrading worms." Available at: www.scientificamerican.com/article/plastic-eating-worms-could-inspire-waste-degrading-tools/

57 "Plastic road." Available at: www.plasticroad.eu/en/. Press release, October 20, 2016, Kws, Wavin and Total Sign a Cooperation Agreement to Develop the Plastic Road. Available at: https://plasticroad.eu/wp-content/uploads/2016/10/Press-Release-20.10.2016.pdf

58 Dykes Paving. "Texas roads made from plastic." Available at: www.dykespaving. com/blog/texas-roads-made-from-plastic/

59 The Drive. "Roads made of recycled plastic are being tested in Britain." Available at: www.thedrive.com/news/9748/roads-made-of-recycled-plastic-are-being-tested-in-britain

60 Earth Untouched. "Plastic roads." Available at: http://earthuntouched.com/plastic-roads-revolutionary-idea/

61 Inspir-action. "Netherlands: Plastic roads to be made from recycled ocean waste." Available at: www.inspiraction.news/en/2016/09/16/netherlands-plastic-roads-to-be-made-from-recycled-ocean-waste/

62 Arch Daily. "This house was built in 5 days using recycled plastic bricks." Available at: www.archdaily.com/869926/this-house-was-built-in-5-days-using-recycled-plastic-bricks

63 Spectrum. "A tower of molten salt will deliver solar power after sunset." Available at: http://spectrum.ieee.org/green-tech/solar/a-tower-of-molten-salt-will-deliver-solar-power-after-sunset

64   Phys.org. "Let there be light: German scientists test 'artificial sun.'" Available at: [[QUERY – PLS SUPPLY WEB ADDRESS?]]

65   Futurism. "The world's largest artificial sun is powering up in Germany." Available at: https://futurism.com/the-worlds-largest-artificial-sun-is-powering-up-in-germany/

66   Anthony King. "Atmospheric limestone dust injection could halt global warming." December 16, 2016. *Chemistry World*. Available at: www.chemistryworld.com/news/atmospheric-limestone-dust-injection-could-halt-global-warming/2500141.article

67   Solar Radiation Management Governance. Available at: www.srmgi.org/

68   Live Science. "Could space mirrors stop global warming?" Available at: www.livescience.com/22202-space-mirrors-global-warming.html

69   Pan-Arctic Ice Ocean Modeling and Assimilation System. "Arctic sea ice volume reanalysis." Available at: http://psc.apl.uw.edu/research/projects/arctic-sea-ice-volume-anomaly/

70   Inside Climate News. "Extreme Arctic melt is raising sea level rise threat." Available at: https://insideclimatenews.org/news/25042017/arctic-sea-ice-climate-change-global-warming-sea-level-rise-ipcc

71   *National Geographic.* "What would the world look like if all of the ice melted?" Available at: www.nationalgeographic.com/magazine/2013/09/rising-seas-ice-melt-new-shoreline-maps/

72   Robin McKie. "Could a € 400bn plan to refreeze the Arctic before the ice melts really work?" *The Guardian*, February 11, 2017. Available at: www.theguardian.com/world/2017/feb/12/plan-to-refreeze-arctic-before-ice-goes-for-good-climate-change; http://onlinelibrary.wiley.com/doi/10.1002/2016EF000410/full

73   National Science Foundation. "Clouds: The wild card of climate change." Available at: www.nsf.gov/news/special_reports/clouds/question.jsp

74   *Architectural Digest.* "Iconic route 66 America's first solar highway." Available at: www.architecturaldigest.com/story/iconic-route-66-americas-first-solar-roadway

75   Arstechnica. "World's first solar road opens in France." Available at: https://arstechnica.com/cars/2016/12/worlds-first-solar-road-opens-in-france/

76   Think progress. "The world's first solar road is producing more energy than expected." Available at: https://thinkprogress.org/the-worlds-first-solar-road-is-producing-more-energy-than-expected-c51540906eb

77   *National Geographic.* "Will we soon be riding on solar roads?" Available at: http://news.nationalgeographic.com/energy/2016/03/160310-will-we-soon-be-riding-on-solar-roads/

78   Inter-press Service. "Solar tents improve nutrition in highlands villages in Bolivia." Available at: www.ipsnews.net/2017/06/solar-tents-improve-nutrition-in-highlands-villages-in-bolivia/

79   Aljazeera. "Solid rain: A possible Mexican solution for drought." Available at: www.aljazeera.com/video/news/2017/03/solid-rain-mexican-solution-drought-170322151958207.html

80   Modern Farmer. "Can powdered water cure droughts? Available at: http://modernfarmer.com/2013/07/powdered-water-just-add-water/

81   CBS News. "Brazil soccer field harnesses player-power". September 11, 2014. Available at: www.cbsnews.com/news/soccer-field-power-players-kinetic-energy-brazil-electricity/

82   Living Future. Available at: https://living-future.org/lbc/

83 *Scientific American*. "Green architecture. What makes a structure a living building?" Available at: www.scientificamerican.com/article/earth-talks-living-building/

84 Perovskite-info. Available at: www.perovskite-info.com/perovskite-solar

85 *Rfi.fr/france*. "Hydrogen-powered catamaran Energy Observer launched in sea at Saint Malo. Available at: http://en.rfi.fr/france/20170415-hydrogen-powered-catamaran-energy-observer-launched-sea-saint-malo

CHAPTER 10

We Are All in the Same Canoe

Prime Minister of Fiji, Frank Bainimarama, as president of COP23, opened the Climate Change meetings in Bonn on November 6. 2017. In closing his welcoming address he said:

> We are all in the same canoe, which is why we have [brought along] *Drua* – a Fijian ocean-going canoe – . . . to remind us of our duty to fill its sail with a collective determination to achieve our mission. So let's make the hard decisions that have to be made for the sake of ourselves and the generations to come. Let's use the next two weeks to get the job done. *Vinaka vakalevu.* Thank you.[1]

The idea here is as follows: To achieve the goal we all must share and for the sake of future generations, we must take up the responsibility—*as a collective duty and with collective determination*—to work together. The immediate goal is *to fill the Drua's sail.*

In other words, the metaphor is not "we're in the same boat, awaiting our fate," but rather, "we are all working together to achieve a shared goal," or, maybe, "all hands on deck." We work together to "fill the sail." Seafarers will immediately know what that means. For the rest of us, it means for all of to work together to fill the sail with wind. In other words, we work together for a common goal. What comes to mind are quite a few synonyms: social contract, reciprocity, egalitarian, participatory society, social compact, communist, democratic society, communitarian, socialist. Call it what you will. I am not precise because the simple idea is that if we don't work together, we are doomed. Which term you choose depends on where you come from and your society's history. Americans may prefer "democratic," Cubans may prefer "communistic." It doesn't matter, since when it gets down to getting the task done, there is no practical difference.

The point that Prime Minister and President Bainimarama is making is that the *entire world—all countries* and *all people everywhere*—must work together, as equals, to slow the planet from overheating, to adopt forms of renewable energy, to respond to emergencies, and to share resources. The biggest obstacle to working together, I argue, is racial inequality. The blacker the population of a country, the poorer the health outcomes. In other words, to work together as equals to keep the planet from overheating we need to level the playing field. Predominately black and predominately white nations need to have equal resources and equal wherewithal to tackle climate change.

Economic inequality is another obstacle. The purpose of the Green Climate Fund is for rich countries to assist poorer countries acquire renewable energy. As I have already mentioned, this is only fair since countries that are now rich are largely those that spewed up the carbon dioxide that lingers in the air today. Economic inequality plays another role as well. When wealth or income is highly concentrated in the hands of a small elite, it is justifiably perceived to be unfair by the non-rich. The rich tend to be irresponsible and selfish. That is true anywhere. It is especially true in America. The richest 10% of American households earns about 28% of the overall income pie. Wealth inequality is more extreme and greater than in other developed countries. The wealthiest 10% of U.S. households have 76% of all the wealth in America. True enough, inequality everywhere has increased under the conditions of globalization, but much more in the U.S.[2] Wealthy people are more often than not indifferent to the conditions. Good citizenship depends on empathy and compassion. And, the more inequality there is in a society, the more distant the rich put themselves from all the rest.

The need for urgency is obvious. Our world is in distress from the extreme weather events caused by climate change—destructive hurricanes, fires, floods, droughts, melting ice, and changes to agriculture that threaten our food security. All are consistent with the science that now tells us that 2016 was a record year for carbon concentrations in the atmosphere.

All over the world, vast numbers of people are suffering, bewildered by the forces ranged against them. Our job as leaders is to respond to that suffering with all the means available to us. This includes our capacity to work together to identify opportunities in the transition we must make.

We must not fail our people. We must do everything we can to make the Paris Agreement work and to advance ambition and support for climate action before 2020.

We must meet our commitments in full, not back away from them, and commit ourselves to the most ambitious target of the Paris Agreement: To cap the global average temperature at 1.5°C over that of the pre-industrial age.

One way we can capture this moment in time is to clarify that never before in human history was it possible for one person or one country to destroy the world. Yes, that would be Trump. Yet, building on universal cooperation—if America rejoins Paris—the world could build on that success. Speculatively and optimistically, if 100% cooperation with respect to Paris is achieved, that can be the platform for future cooperation—say, to achieve universal food security, end homelessness, empower women, achieve universal education, provide opportunities for refugees, end wars.

NOTES

1   Frank Bainimarama, "We must not fail our people." President's opening speech at COP23. Available at: https://cop23.com.fj/fijian-prime-minister-cop23-president-remarks-assuming-presidency-cop23/

2   There have been many careful studies of income and wealth inequalities. See, for example, Emmanuel Saez, "Income and wealth inequality: Evidence and policy implications," *Contemporary Economic Policy*, 35: 7–25, January 2017. https://eml.berkeley.edu/~saez/SaezCEP2017.pdf

CHAPTER 11

Concluding Remarks

I wish to stress four major points in this final chapter. First, warming is accelerating much faster than scientists had predicted. Second, "there is no going back," or "going back entails very risky geo-engineering." Third, I will explore the possibility that Americans do not "get" the United Nations framework, which, although it is a legal framework, is based on commitments to cooperate and a shared trust that other countries will do the same. We are, after all, in the same canoe. Fourth, I will consider the possibility that Americans do not understand the seriousness of climate change. Trump does not.

FASTER ACCELERATION OF WARMING THAN PREDICTED

The following statement appears in an article published in the *Proceedings of the National Academy of Science*, which is truly frightening: *Do read it slowly and carefully.*

> Any exceedence of 35°C [95°F] for extended periods should induce hyperthermia in humans and other mammals, as dissipation of metabolic heat becomes impossible. *While this never happens now, it would begin to occur with global-mean warming of about 7°C [44.6 °F], calling the habitability of some regions into question. With 11–12°C [51–52°F] warming, such regions would spread to encompass the majority of the human population as currently distributed.*[1]

The statement below, also scary, is on the website of National Aeronautics and Space Administration (NASA), and NASA's websites remain, amazingly, on the Web, whereas those referring to climate change on the websites of the Environmental Protection Agency have come down.

It takes decades to centuries for earth to fully react to increases in greenhouse gases. Carbon dioxide, among other greenhouse gases, will remain in the atmosphere long after emissions are reduced, contributing to continuing warming. In addition, as earth has warmed, much of the excess energy has gone into heating the upper layers of the ocean. *Like a hot-water bottle on a cold night, the heated ocean will continue warming the lower atmosphere well after greenhouse gases have stopped increasing.*[2]

The following is a quotation from the journal *Science:* "Global temperatures could rise nearly 5°C by the end of the century under the UN Intergovernmental Panel on Climate Change's steepest prediction for greenhouse-gas concentrations."[3]

GEO-ENGINEERING IS TOO RISKY AND/OR NOT WELL EXPLAINED TO THE PUBLIC

I mention this because geo-engineering is often viewed as a quick fix. In Chapter 9, I summarized several of the very ambitious proposals to slow or reverse warming, including, for example, spraying sulfur into the atmosphere or blocking the sun with mirrors. However, these proposals have been scuttled because they are too risky or would be impossible to reverse.

THE UNITED NATIONS FRAMEWORK AND AMERICAN VIEWPOINTS

Americans do not "get" the United Nations generally because the United States and the United Nations rest on different organizational and language frameworks. The United States relies to an extraordinary degree on laws, and its constitution is defined in terms of law. For this reason, Americans may not easily relate to UN institutions and treaties (such as the Paris Agreement), and it has not ratified human rights and other UN treaties. The United Nations stresses equality, human dignity and a shared sense of justice rather than laws. This is Article 1 of the UN Charter.

The Purposes of the United Nations are:

1. To maintain international peace and security, and to that end: to take effective collective measures for the prevention and removal of threats

to the peace, and for the suppression of acts of aggression or other breaches of the peace, and to bring about by peaceful means, and in conformity with the principles of justice and international law, adjustment or settlement of international disputes or situations that might lead to a breach of the peace.

2.   To develop friendly relations among nations based on respect for the principle of equal rights and self-determination of peoples, and to take other appropriate measures to strengthen universal peace.

3.   To achieve international cooperation in solving international problems of an economic, social, cultural, or humanitarian character, and in promoting and encouraging respect for human rights and for fundamental freedoms for all without distinction as to race, sex, language, or religion.

4.   To be a centre for harmonizing the actions of nations in the attainment of these common ends.

Note that the word "law" or "laws" appears in the U.S. Constitution 23 times and in the UN Charter twice. The U.S. Constitution emphasizes order and laws, whereas the UN Charter emphasizes world peace, collaboration, and harmony. It is possible that Americans do not "get" the idea of collaboration because collaboration does not match America's conception that an international order is based on the pursuit of national self-interest, within a framework of laws, and not with a spirit of cooperation.

DO AMERICANS UNDERSTAND THE PARIS AGREEMENT?

Below is a paragraph from the Preamble of the Paris Agreement:

> Acknowledging that climate change is a common concern of humankind, Parties should, when taking action to address climate change, respect, promote and consider their respective obligations on human rights, the right to health, the rights of indigenous peoples, local communities, migrants, children, persons with disabilities and people in vulnerable situations and the right to development, as well as gender equality, empowerment of women and intergenerational equity.[4]

Note that the above paragraph—and the entire treaty—does not include the word, "law" or "laws," although it is a legal treaty. The premise instead is "we share a common humanity and will do our best to protect one another." This contrasts with any particular American program or treaty, the American legal framework, and American laws, in general. We are a nation of individualists, or we could say, a nation of individualists who possess self-interest. (Only laws will keep us in check.) It is useful to point out that the term "law" or "laws" appears in the U.S. Constitution 23 times and in the UN Charter twice. We could say that Americans emphasize "right and wrong," or "according to the books," whereas the international community tends to emphasize "how do we cooperate to get it done?" or "let's get together and do it." I also imply that Americans do not realize that important UN treaties, laws and agreements are legal documents—including, for example, human rights treaties, the UN Charter, the Kyoto Protocol, Convention Related to the Status of Refugees, and the Paris Agreement.

Another consideration is that Americans, as reported in Chapter 1 (according to an international Pew Research survey), are least likely to believe that climate change is taking place, which may account for why Trump could so easily withdraw from the Paris Agreement. The consequences will be horrific for the entire world. The seas will rise; many places will be so hot they will be uninhabitable; storms will be fierce; the Arctic will melt; many animals and fish will become extinct; wells will dry up; the corals will die; small island states will disappear; and there will be millions of climate refugees. Yes, it is a crime against humanity. But let us hope that the U.S. learns quickly—very quickly—the importance of international cooperation.

NOTES

1   Steven C. Sherwood and Matthew Huber, "An adaptability limit to climate change due to heat stress." PNAS 2010, May, 107(21): 9552–9555. Available at: https://doi.org/10.1073/pnas.0913352107 (emphasis added).

2   National Aeronautics and Space Administration. "Global warming." Available at: https://earthobservatory.nasa.gov/Features/GlobalWarming/page5.php (emphasis added).

3   Patrick T. Brown and Ken Caldeira, "Greater future global warming inferred from eaarth's recent energy budget." *Nature*, 552: 45–50, December 7, 2017. doi:10.1038/nature24672

4   Paris Agreement. Available at: http://unfccc.int/files/essential_background/convention/application/pdf/english_paris_agreement.pdf

Online Resources on Climate Change

The following are brief descriptions and websites of organizations/social movements that have e-newsletters devoted to climate change. There is no charge for subscriptions to newsletters.

350.0rg

350.org is building the global grassroots climate movement that can hold our leaders accountable to science and justice.

Available at: https://350.org/about/

Center for Climate and Energy Solutions: C2ES

Our mission is to advance strong policy and action to reduce greenhouse gas emissions, promote clean energy, and strengthen resilience to climate impacts. A key objective is a national market-based program to reduce emissions cost-effectively. We believe a sound climate strategy is essential to ensure a strong, sustainable economy.

Available at: www.c2es.org/about

Center for Coastal Studies

To conduct scientific research with emphasis on marine mammals of the western North Atlantic and on the coastal and marine habitats and resources of the Gulf of Maine; to promote stewardship of coastal and marine ecosystems; to conduct educational activities and to provide educational resources that encourage the responsible use and conservation of coastal and marine ecosystems; and to collaborate with other institutions and individuals whenever possible to advance the Center's mission.

Available at: http://coastalstudies.org/

Citizens Climate Lobby

We exist to create the political will for climate solutions by enabling individual breakthroughs in the exercise of personal and political power.

Available at: https://citizensclimatelobby.org/

Climate Action Network International (CAN)

CAN is a network of NGOs working on climate change from around the world. These members are autonomous and independent. Many of these members have their own forms of organization and their own national or regional rules.

Available at: www.climatenetwork.org/

Climate Analytics

Climate Analytics is a non-profit climate science and policy institute based in Berlin, Germany with offices in New York, USA and Lomé, Togo, which brings together interdisciplinary expertise in the scientific and policy aspects of climate change. Our mission is to synthesize and advance scientific knowledge in the area of climate change and provide support and capacity building to stakeholders. By linking scientific and policy analysis, we provide state-of-the-art solutions to global and national climate change policy challenges.

Available at: http://climateanalytics.org/

Climate Central

An independent organization of leading scientists and journalists researching and reporting the facts about our changing climate and its impact on the public. Distinct web pages with links.

Available at: www.climatecentral.org/
Surging seas: Available at: http://sealevel.climatecentral.org/

States of change: Available at: www.climatecentral.org/states-of-change#/nation

Climate Home (Climate Change News)

Climate Home has a network of world-class correspondents and reporters based in Delhi, Brussels, the Amazon, Brisbane, Washington, DC, Nairobi, and Addis Ababa. Our London-based editorial team coordinates deep reporting on the political, economic, social and natural impacts of climate change. Our coverage of the UN climate talks is essential reading.

Available at: www.climatechangenews.com/

Daily Climate

The Daily Climate is an independent media organization working to increase public understanding of climate change, including its scope and scale, potential solutions and the political processes that impede or advance them. The Daily Climate does not espouse a political point of view on the news, but instead reports the issue to the best of our ability. Editorial integrity is the foundation of our mission. Our reporting, editing and publishing adheres to the highest standards of journalism, including honesty, accuracy, balance and objectivity.

Available at: www.dailyclimate.org/frontpage/

Drawdown

Drawdown is facilitating a broad coalition of researchers, scientists, graduate students, Ph.D.s, post-docs, policy makers, business leaders, and activists to assemble and present the best available information on climate solutions in order to describe their beneficial financial, social and environmental impact over the next thirty years.

Available at: www.drawdown.org/about

Environmental Defense Fund

Climate change threatens our future. Yet we have many reasons for hope. To stop the rise of climate pollution, while growing the economy, we've zeroed in on solutions with the biggest impact.

Available at: www.edf.org/climate

Friends of the Earth International

Friends of the Earth strives for a more healthy and just world. We understand that the challenges facing our planet call for more than half-measures, so we push for the reforms that are needed, not merely the ones that are politically easy. Sometimes, this involves speaking uncomfortable truths to power and demanding more than people think is possible. It's hard work. But the pressures facing our planet and its people are too important for us to compromise.

Available at: https://foe.org/about-us/friends-of-the-earth-international/

Greenpeace

Greenpeace exists because this fragile earth deserves a voice. It needs solutions. It needs change. It needs action. Greenpeace is an independent global campaigning organization that acts to change attitudes and behavior, to protect and conserve the environment and to promote peace.

Available at: www.greenpeace.org/international/en/

Inside Climate News

Inside Climate News is an independent, not-for-profit, non-partisan news organization that covers clean energy, carbon energy, nuclear energy and environmental science, plus the territory in between where law, policy, and public opinion are shaped.

Available at: https://insideclimatenews.org/

La Via Campesina

La Via Campesina is an international movement bringing together millions of peasants, small and medium-size farmers, landless people, rural women and youth, indigenous people, migrants and agricultural workers from around the world. Built on a strong sense of unity and solidarity between these groups, it defends peasant agriculture for food sovereignty as a way to promote social justice and dignity, and strongly opposes corporate-driven agriculture that destroys social relations and nature.

Available at: https://viacampesina.org/en/

Microgrid Knowledge

Within microgrids are one or more kinds of distributed energy (solar panels, wind turbines, combined heat and power, generators) that produce its power. In addition, many newer microgrids contain energy storage, typically from batteries. Some also now have electric-vehicle charging stations. Interconnected to nearby buildings, the microgrid provides electricity, and possibly heat and cooling for its customers, delivered via sophisticated software and control systems.

Available at: https://microgridknowledge.com/

Modeling European Agriculture with Climate Change for Food Security (MACSUR)

The knowledge hub MACSUR aims at improving modeling methodologies and sound case study applications leading to improved assessments of climate change on European agriculture.

Available at: https://macsur.eu/

Mothers Out Front

The world is starting to understand that we must act now to address climate change. No matter where we each stand on the political spectrum, we are united in our belief that our work is more important now than ever. We have a short but real window of time to act. We have the knowledge, skills, and much of the technology we need to keep the earth's temperature from rising to catastrophic levels.

Available at: www.mothersoutfront.org/

National Academies of Sciences, Engineering, and Medicine

At a time when responding to climate change is one of the nation's most complex and urgent endeavors, reports and convening activities of the National Academies of Sciences, Engineering, and Medicine (NASEM) provide objective guidance in support of policy- and decision-making. The NASEM gathers the nation's top scientific and technical experts to address specific questions through rigorous, independent, and evidence-based processes. Findings and recommendations from the Academies have helped the nation move forward in understanding and addressing climate change.

Available at: https://nas-sites.org/americasclimatechoices/about-climate-change-at-the-nasem/

National Aeronautics and Space Administration (NASA)

NASA Global Climate Change. Vital Signs of the Planet (also NASA's Climate Change Newsletter). Global Climate Change website is produced by the Earth Science Communications Team at NASA's Jet Propulsion Laboratory/Caltech.

Available at: https://climate.nasa.gov/

National Resources Defense Council

NRDC works to safeguard the earth—its people, its plants and animals, and the natural systems on which all life depends.

Available at: www.nrdc.org/

Rainforest Action Network

Rainforest Action Network preserves forests, protects the climate and upholds human rights by challenging corporate power and systemic injustice through frontline partnerships and strategic campaigns.

Available at: www.ran.org/

Reclaim Power

Toward transforming energy systems and keeping global temperature below 1.5°C. Stop new dirty energy projects. End public handouts to dirty energy. Divest from fossil fuels. Redirect funds and pursue a swift and just transition to democratic, pro-poor, 100% renewable and clean energy systems for people and communities. Stop excessive energy consumption by corporations and elites. Ensure universal energy access for basic needs of people and communities.

Available at: www.reclaimpower2017.net/

Renewable Energy World

The World's #1 Renewable Energy Network for News, Information, and Companies. Several newsletters: Solar, wind, energy

Available at: www.renewableenergyworld.com/index.html

Scientific American

Comprehensive and clear reports latest news and reports on science.

Available at: www.scientificamerican.com/

Sustainable Development Goals
(United Nations)

In 2015, countries adopted the 2030 Agenda for Sustainable Development and its 17 Sustainable Development Goals. In 2016, the Paris Agreement on climate change entered into force, addressing the need to limit the rise of global temperatures. Explore this site to find out more about the efforts of the UN and its partners to build a better world with no one left behind.

Available at: www.un.org/sustainabledevelopment/
www.drawdown.org/about

Union of Concerned Scientists

UCS was founded in 1969 by scientists and students at the Massachusetts Institute of Technology. That year, the Vietnam War was at its height and Cleveland's heavily polluted Cuyahoga River had caught fire. Appalled at how the U.S. government was misusing science, the UCS founders drafted a statement calling for scientific research to be directed away from military technologies and toward solving pressing environmental and social problems. We remain true to that founding vision. Throughout our history, UCS has followed the example set by the scientific community: We share information, seek the truth, and let our findings guide our conclusions.

Available at: www.ucsusa.org/

United Nations Framework Convention on Climate Change

Newsroom. Up-to-date announcements, meetings, progress reports. Updated daily.

Available at: http://newsroom.unfccc.int/

Woods Hole Oceanographic Institution

Woods Hole Oceanographic Institution is the world's leading independent non-profit organization dedicated to ocean research, exploration, and education. Our scientists and engineers push the boundaries of knowledge about the ocean to reveal its impacts on our planet and our lives.

Available at: www.whoi.edu/

Woods Hole Research Center

WHRC is an independent research institute where scientists investigate the causes and effects of climate change to identify and implement opportunities for conservation, restoration and economic development around the globe. In June 2017, WHRC was ranked as the top independent climate change think tank in the world for the fourth year in a row.

Available at: http://whrc.org/
https://citizensclimatelobby.org/

World Resources Institute

WRI's mission is to move human society to live in ways that protect earth's environment and its capacity to provide for the needs and aspirations of current and future generations.

Available at: www.wri.org/

Yale Climate Connections

Yale Climate Connections is a nonpartisan, multimedia service providing daily broadcast radio programming and original Web-based reporting, commentary and analysis on the issue of climate change, one of the greatest challenges confronting modern society.

Available at: www.yaleclimateconnections.org/
https://microgridknowledge.com/
www.dailyclimate.org/frontpage/

Reducing Our Own Carbon Footprint

France's Minister of the Environment, Nicolas Hulot, announced on July 5, 2017 that France will ban petrol cars by 2040. He also said that France will stop producing power from coal by 2022—now 5% of the total—and that the country will reduce the proportion of its power from nuclear energy to 50% by 2025, from the current 75%. This followed Volvo's announcement on July 4 that the company would stop producing petrol and diesel cars by 2019.[1] After that, government officials announced that the UK would disallow gas-driven cars after 2040. And, in fact, European leaders have agreed to end fossil fuel subsidies by 2025. In that context, one might suspect that Trump's announcement to pull out of the Paris Agreement and then his subsequent rollback of domestic climate change programs (as noted in Chapter 8) would lead to such despair in America that climate change initiatives in the U.S. would cease. As it turns out, this has not been the case at all. At the June 2017 conference of U.S. city mayors more than 250 signed a resolution to set a target of 2035 for their cities to use 100% renewables. And then there was an international movement with mayors of more than 7,400 cities voting to work together to slow climate change.[2] Also, in the United States, California has taken a leadership role, headed by Governor Jerry Brown.

But, quite surprisingly, there is a grassroots social movement in America to reduce our carbon footprints. Yes, like most social movements, it is based on collectivized and aggregated individual efforts, and once it initially takes off in a community, more and more jump on. In this appendix I simply list the many possibilities.

I will use the term "carbon footprint" casually, as is generally the case, but noting that it has a technical meaning—namely, termed CO_{2e}, it is the carbon dioxide (CO_2) equivalent of all the different greenhouse gases.[3] So "carbon footprint" is shorthand for all the different greenhouse gases that contribute to global warming. That way, a carbon footprint consisting of many different greenhouse gases can be expressed as a single number.

The average American has a bigger carbon footprint that the average European. More precisely, the average estimated U.S. carbon footprint is 16.5 tons per year and that for the European of 6.8 tons per year. The per capita average for the world as a whole is even lower at 5 tons of CO_{2e} per year. There are also indications that many more Americans are unaware of climate change than Europeans, and that climate change skepticism and denial are higher in the U.S. than in Europe, and especially in Latin America.[4]

To be clear, reducing their carbon footprint is not a project to be pursued by millions—even billions—of individuals, acting as individuals. It can only be a *collective project*, grounded in the grassroots with everyone chipping in, while all being motivated by cosmopolitan global determination. This is where one person's ideas and insights spark others, and in turn, each sparks more so that there is a contagion process and the world's peoples share ideas and proposals. We all have a stake. I will start with very modest ways of addressing climate change and reducing CO_2, and then discuss more ambitious approaches.

AT HOME

Pay your bills online and stop paper bank statements.

Turn off lights when you don't need them.

Air dry your hair and clothes.

Recycle paper, plastic, glass, aluminum.

Plug air leads in windows and doors.

Lower thermostat in the winter, make it higher in the summer.

Carpets and rugs keep the house warmer than bare floors.

Use cloth or environmentally friendly disposable diapers.

Use cardboard matches, which don't require petroleum (unlike plastic, gas-filled lighters).

Take a shower rather than a bath (which requires more water).

Dry clothes on a rack, not in the dryer.

Donate what you don't use. Local charities will give your gently used clothes, books and furniture a new life.

Eliminate junk mail.[5]

In winter you can open your curtains during the day and close them at night, while in summer you can leave them open at night and close them during the day. Doing this, it is possible to save up to 75% on your bill.

Reading news online rather than in print saves paper and printing costs.

Ceiling fans instead of air conditioning save CO_2 and money.

A programmable thermostat costs about $50 or less and will save you that much or more in the first year.

Weather stripping and caulking costs almost nothing while reducing your energy use, reducing drafts, and improving comfort.

Compact Fluorescent Lightbulbs (CFLs) have that cool curly shape and save more than two-thirds of the energy of a regular incandescent bulb. Each bulb can save $40 or more over its lifetime. LEDs are even more efficient to run than CFL bulbs and are just as affordable. In addition, LED lights contain no harmful materials and provide the same amount of light as a 60-watt incandescent, using up to 85% less power. This helps you save on your electricity bill without compromising on the amount of light.

Weather stripping, caulking, and insulation work together to save you energy, improve the comfort of your home, make it quieter and help you save money.

Install water-conserving showerheads and toilets in your bathroom. To save even more water, turn the faucet off when brushing or shaving. These simple changes and steps can save many thousands of gallons of water annually.

Use recycled paper for sending cards, letters, or wrapping gifts.

Don't use black trash bags. Because of the black pigmentation, these trash bags cannot be recycled. Bags certified by BPI are compostable.

Wash and rinse clothes with cold water to save energy by not using hot-water heater.

If you have windows you can open, use them to intelligently save energy.

Turn lights off—only use the lights you need.

Install occupancy sensors for room lights.

Lower the amount of energy used to pump, treat, and heat water by washing your car less often, installing drip irrigation so that plants receive only what they need, and making water-efficient choices when purchasing shower heads, faucet heads, toilets, dishwashers and washing machines.

Buy wood that is certified by the Forest Stewardship Council that certifies that wood comes from responsibly managed forests. This means that the logging in those forests is being monitored to prevent deforestation and maintain biodiversity.[6]

Move closer to work.

Use gift wrapping with seeds; the paper can be planted and flowers will grow.

Having one fewer child on average (for developed countries) will save 58.6 tons of CO_{2e} per year.[7]

Use CFL or LED lights.

OUTDOORS

Use walk-behind lawnmower.

Plant a tree[8] (trees absorb carbon dioxide).

Paint your house roof green to reduce the energy needed for summer heating and winter cooling, and help rainwater retention to reduce runoff problems. Another plus is that they create an excellent place for birds to nest within urban environments. Therefore, green roofs have financial and environmental benefits suggesting that cities and countries should make it easy for residents and businesses to adopt them. France, in fact, requires the roofs of all new buildings to be green (or solar).[9]

Shovel snow manually.

COMPUTER AND APPLIANCES

Upgrade appliances with greater efficiency.

Unplug unused appliances.

Unplug and disconnect cell phone charger.

If you're using your computer, you may not need your office lights on too.

Appliances/windows/doors/skylights/entire houses/computers: In the United States, buy what is marked Energy Star, which are certified, after testing, as saving energy.[10]

Plug appliances into a power strip (including computer) and turn off completely when not in use.

Use energy-efficient appliances and lightbulbs.

Laptops use up to 80% less energy than a desktop.

SHOPPING

Shop vintage. Brand new isn't necessarily best. See what you can repurpose from second-hand shops.

Buy paper straws and use them instead of plastic.

Shop local. Supporting neighborhood businesses keeps people employed and helps prevent trucks from driving long distances. Shipping burns fuel. A 5-pound package shipped by air across the country creates 12 pounds of CO_2 (3½ pounds if shipped by truck).

Bring your own bag when you shop. Pass on the plastic bag and start carrying your own reusable totes.

FOOD

Consume no or less meat; eat poultry and fish and more vegetables. (Fewer resources are used to produce vegetables). If you're already a vegetarian, you save at least 3,000 pounds of CO_2 per year compared to meat eaters. If you're not a vegetarian, just increase the number of vegetarian meals you eat each week by one or two. Also, poultry is less greenhouse-gas intensive than beef.

Eat less or no beef (cows produce methane).

Freeze fresh produce and leftovers to eat later.

Compost, which reduces climate impact.

Bring your own glass straw to the restaurant; better yet, use a paper straw.

Lawns are not necessary. Grow vegetables instead.

Take fewer napkins. You don't need a handful of napkins to eat your takeout. Take just what you need.

On average, 0.5 lbs. of beef creates more carbon dioxide than driving for 5 miles. On average, the emission created from providing a half a pound of beef steaks has the same harmful effects as driving 9.81 miles.

Your microwave is more energy efficient than your electric stove.

Bring your lunch, or walk to the local eatery instead of driving.

Don't waste food. Mom was right. About one-quarter of all the food prepared annually in the U.S., for example, gets tossed, producing methane in landfills as well as carbon emissions from transporting wasted food.

Buy locally produced food. About 13% of U.S. greenhouse gas emissions are from the production and transport of food. Transporting food requires petroleum-based fuels, and many fertilizers are also fossil fuel-based.

Buy "funny fruit"—many fruits and vegetables are thrown out because their size, shape, or color are not "right." Buying these perfectly good funny fruit, at the farmer's market or elsewhere, utilizes food that might otherwise go to waste.

Use a refillable water bottle and coffee cup. Cut down on waste and maybe even save money at the coffee shop.

TRANSPORTATION

Cycle, walk, or take public transport.[11]

Maintain your car. A well-tuned car will emit fewer toxic fumes.

Buy and drive an electric car that you plug into an outlet from your solar roof.[12] There are other alternatives: gas–electric hybrids, plug-in hybrids, electric vehicles, and vehicles to run on ethanol, biodiesel, propane, liquefied natural gas, and solar.[13]

You could save more than a ton of CO_2 per year by: Accelerating slowly and smoothly; driving the speed limit; maintaining a steady speed; anticipating your stops and starts; keeping your car tuned up and running efficiently; replacing your air, oil and fuel filters according to schedule. Keep your tires properly inflated (just this can save 400–700 pounds of CO_2 per year).

Shift gears sooner. If you drive a manual, shift into a higher gear as soon as you can. On most cars this would mean before 2,500 rpm, but on diesel it would be before 2,000 rpm.

For, say, a cross-continental trip, take a train rather than fly, and fly rather than drive.[14]

Carpooling is a great way to help reduce carbon emissions while also providing you with networking opportunities.

Electric cars: The carbon footprint of your electric car depends on the power source you use. If your electricity comes from a coal-fired plant, then charging your electric car from the grid is no better than powering your car with gasoline. In the U.S., on average, the combination of electricity sources makes the electric car about as efficient as a hybrid.

A FEW MORE SUGGESTIONS

Unwanted clothes: The first solution is to donate gently used clothes to a charity organization. The second solution is to bury them, but don't send them to the dump. The clothing industry is the second most polluting business after oil, contributing to the accumulation of CO_2 and requiring a tremendous amount of water.[15] But, in fact, much clothing—that is, clothing made from cotton or hemp—is biodegradable and can simply be tossed into the garden, with apple peels, left-over veggies, and scrunched-up paper. There are some items, however, that shouldn't be tossed into the compost heap, such as dyes, buttons, and polyester thread and tags. But a Swiss manufacturer, Freitag, has designed a new line of clothing that's safe for your garden or your city's compost bins. Their company sells 100% biodegradable clothes, using nuts instead of buttons, and reusable zippers and buckles.[16]

Inform your air carrier that you support voluntary compliance with CORSIA (Carbon Offsetting and Reduction Scheme for International Aviation), an agreement that supports airlines curbing greenhouse gases through offsetting (see below).[17]

A carbon offset is a way to compensate for your emissions by funding an equivalent carbon dioxide saving elsewhere. Our everyday actions, at home and at work, consume energy and produce carbon emissions, such as driving, flying and heating buildings. Carbon offsetting is used to balance these emissions by helping to pay for emission savings in other parts of the world.[18] 1 carbon offset = 1 metric ton of CO_{2e} kept from the atmosphere; 1 carbon offset represents the reduction of greenhouse gases equal to 1 metric ton (or 2,205 pounds) of carbon dioxide equivalent (CO_{2e}). Terrapas sells them at $4.99 per 1,000 pounds.[19]

CONCLUSIONS

Americans can recognize that the U.S. will cause egregious harm to the entire world (and that includes the United States) because of Trump's decision to withdraw from the Paris Agreement. This decision was counter to decades of research, international meetings, and now the global consensus. America stands alone, petulant and seemingly stupid. Without being a party to an international treaty to which all other countries are a party and that agree to end fossil fuels ideally by 2030, the U.S. has set itself up to destroy the world. We must do our part—individually and collectively—taking all steps possible to reduce our carbon footprint. We can overcome this destructive act by individually and collectively engaging in activities that slow warming. As I stress throughout, we stand in solidarity with the rest of the world and affirm the clarion call of COP23 and its Fijian president: "We are all in the same canoe."

NOTES

1   Angelique Chrisafis and Adam Vaughan. "France to ban sales of petrol and diesel cars by 2040." *The Guardian*, July 6, 2017. Available at: www.theguardian.com/business/2017/jul/06/france-ban-petrol-diesel-cars-2040-emmanuel-macron-volvo. Adam Vaughan, "All Volvo cars to be electric or hybrid." *The Guardian*, July 5, 2017. Available at: www.theguardian.com/business/2017/jul/05/volvo-cars-electric-hybrid-2019

2   Lizette Alvarez, "Mayors, sidestepping Trump, vow fill void on climate change." *New York Times*. June 26, 2017. Available at: www.nytimes.com/2017/06/26/us/mayors-trump-climate-change.html?_r=0; Daniel Boffey, "Mayors of 7,400 cities vow to meet Obama's climate commitments." *The Guardian*, June 28, 2017. Available at: www.theguardian.com/environment/2017/jun/28/global-covenant-mayors-cities-vow-to-meet-obama-climate-commitments

3   *The Guardian*. "What are CO_2 and global warming potential (GWP)?" April 27, 2011. Available at: www.theguardian.com/environment/2011/apr/27/co2e-global-warming-potential

4   Pew Research Center. What the world thinks about in 7 charts. April 18, 2016. Available at: www.pewresearch.org/fact-tank/2016/04/18/what-the-world-thinks-about-climate-change-in-7-charts/

5   In the United States, one can do this through Data and Marketing Association. Available at: https://thedma.org/accountability/dma-choice/

6   Forest Stewardship Council (FSC). Certification. Available at: https://us.fsc.org/en-us/certification

7   Seth Wynes and Kimberly A. Nicholas. "The climate mitigation gap: Education and government recommendations miss the most effective individual actions." *Environmental Research Letters*, July 12, 2017. Available at: http://iopscience.iop.org/article/10.1088/1748-9326/aa7541/meta

8   Onetreeplanted.org. Carbon footprint facts and statistics everyone should know. Available at: https://onetreeplanted.org/blogs/news/13062461-carbon-footprint-facts-and-statistics-everyone-should-know

9   AE News. "France passes green rooftop law." Available at: www.alternative-energy-news.info/france-green-rooftop-law/. Inhabitat. "France requires all new buildings to have green roofs or solar panels." Available at: http://inhabitat.com/france-requires-all-new-buildings-to-have-green-roofs-or-solar-panels/

10   Energystar. Available at: www.energystar.gov/

11   According to the Union of Concerned Scientists. "Collectively, cars and trucks account for nearly one-fifth of all U.S. emissions, emitting around 24 pounds of carbon dioxide and other global-warming gases for every gallon of gas. About five pounds comes from the extraction, production, and delivery of the fuel, while the

great bulk of heat-trapping emissions—more than 19 pounds per gallon—comes right out of a car's tailpipe. Available at: www.ucsusa.org/clean-vehicles/car-emissions-and-global-warming#.WVvb71GQyUk

12 Electric cars reassure their owners that they are reducing or eliminating emissions, but that is not the case. The source of power for the electric vehicle is of course, the power grid and the U.S. gets about a third of its electricity from coal-fired power. See Anrica Deb. "Why electric cars are only as clean as their power supply." December 8, 2016. Available at: www.theguardian.com/environment/2016/dec/08/electric-car-emissions-climate-change

13 Fortune. "10 Alternatives to the gasoline powered engine." Available at: http://fortune.com/2013/11/01/10-alternatives-to-the-gasoline-powered-engine/

14 National Public Radio. What's greener, flying or driving? Available at: www.npr.org/templates/story/story.php?storyId=15160745

15 "Gather and see." Available at: www.gatherandsee.com/the-gatherer/fashion-and-climate-change/

16 Fast Company. "When these clothes wear out you can throw them in your compost bin." Available at: www.fastcompany.com/3040693/when-these-clothes-wear-out-you-can-throw-them-in-your-compost-bin; "1 million women. How to compost fabrics." September 21, 2016. Available at: www.1millionwomen.com.au/blog/how-compost-fabrics/

17 In 2016, member states of the UN International Civil Aviation Organization approved an agreement to curb greenhouse gas through offsetting International Civil Aviation Organization. The new system will be voluntary until 2027, but dozens of countries, including the world's two largest emitters, the U.S. and China, have promised to join at its outset in 2020. The plan—Carbon Offsetting and Reduction Scheme for International Aviation (CORSIA)—will begin with a pilot phase from 2021 through 2023, followed by a first phase, from 2024 through 2026. Participation in both of these early stages will be voluntary and the next phase (from 2027 to 2035) would see all states on board. Some exemptions were accepted for Least Developed Countries (LDCs), Small Island Developing States (SIDS), Landlocked Developing Countries (LLDCs), and states with very low levels of international aviation activity Historic agreement reached to mitigate international aviation emissions. Available at: www.icao.int/Newsroom/Pages/Historic-agreement-reached-to-mitigate-international-aviation-emissions.aspx

18 Duncan Clark, "A complete guide to carbon offsetting." *The Guardian.* September 16, 2011. Available at: www.theguardian.com/environment/2011/sep/16/carbon-offset-projects-carbon-emissions

19 Terrapass. Available at: www.terrapass.com/product/individuals-families. Carbon footprint. Available at: www.carbonfootprint.com/carbonoffset.html

Index

Note: page references in **bold** indicate tables; *italics* indicate figures; 'n' indicates chapter notes

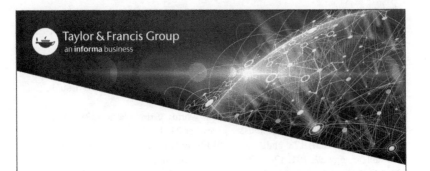